welding skills workbook

Second Edition

AMERICAN TECHNICAL PUBLISHERS, INC.
HOMEWOOD, ILLINOIS 60430

Jonathan F. Gosse

2 3 4 5 6 7 8 9 – 96 – 9 8 7 6 5 4 3 2 1

Printed in the United States of America

ISBN 0-8269-3008-5

CONTENTS

REVIEW QUESTIONS

TECH-CHEKS

SECTION ACTIVITIES

SECTION EXAMS

KNOW YOUR WELDING SYMBOLS

INTRODUCTION

Welding Skills Workbook, 2nd Edition covers material presented in the corresponding chapter of *Welding Skills*, 2nd Edition. The workbook includes Review Questions, Know Your Welding Symbols, Tech-Cheks, Section Activities, and Section Exams.

REVIEW QUESTIONS

Read the assigned chapter in the text. Without using the text, answer the review questions in the corresponding workbook chapter. After answering the review questions, check your answers using the text. Identify and correct any wrong answers. Your instructor will have specific directions for the next assigned activity.

KNOW YOUR WELDING SYMBOLS

Fillet Weld Arrow Side

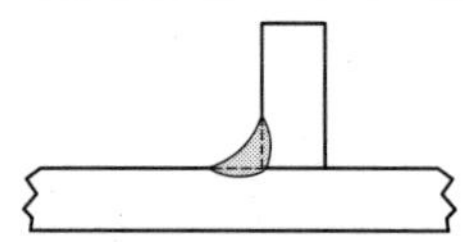

DESIRED WELD

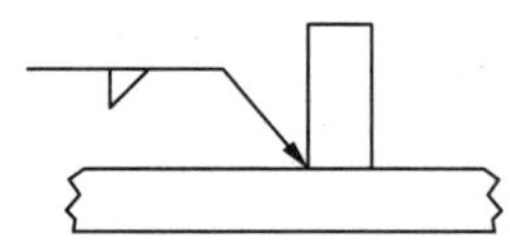

SYMBOL

KNOW YOUR WELDING SYMBOLS

At the bottom of the first page of each new Review Questions chapter, a new welding symbol and its meaning are introduced. This allows progressive learning of common welding symbols used, which are described in Chapter 35. Questions relating to the welding symbols introduced in each chapter and previous chapters are included in the Review Questions and Tech-Cheks.

TECH-CHEKS AND SECTION EXAMS

Tech-Cheks follow the Review Questions, and Section Exams are at the back of the workbook. Your scores on these tests indicate how well you understand the material. Before starting a Tech-Chek or a Section Exam, study the corresponding Review Questions and chapters in the textbook.

SECTION ACTIVITIES

Section Activities follow the Tech-Cheks and feature short-answer questions corresponding to topics covered within the chapters of a section. Welding exercises contained in the Section Activities include text and other related exercises.

SAFETY

Before beginning any exercise, secure permission from the instructor. Observe all safety precautions as stated in the text and as specified by your instructor. Under no circumstance should any machine or equipment be used without proper authorization.

HOW TO ANSWER QUESTIONS

Without using the text, carefully answer the questions in the space provided. Types of questions used in the workbook include True-False, Multiple Choice, and Matching. Questions are answered in the following manner:

For True-False questions, circle T if the statement is true and F if the statement is false.

True-False

(T) F **1.** Eye protection should be worn when welding mild steel.

(T) F **2.** An accident must always be reported no matter how slight it may be.

T (F) **3.** Oxygen should be used to ventilate a closed container before it is welded.

For Multiple Choice questions, place the letter of the correct answer in the blank next to the question.

Multiple Choice

__D__ **1.** When welding mild steel, _________ should be worn to prevent injury to the welder.

A. safety glasses
B. fire resistant clothing
C. welding gloves
D. all of the above
E. none of the above

__C__ **2.** _________ properties refer to the behavior of metals under applied loads.

A. Physical
B. Chemical
C. Mechanical
D. Ferritic
E. all of the above

__B__ **3.** When welding _________, a respirator is required to protect the welder from toxic fumes.

A. aluminum
B. brass
C. steel
D. all of the above
E. none of the above

For matching questions, select the correct answer for each number. Place the letter of the answer in the blank next to the question.

Matching

__B__ **1.** _________ can be harmful to eyes and skin.

__A__ **2.** A welding machine must have a proper _________ to prevent electrical shock.

__E__ **3.** Operating with currents above the rated cable capacity causes _________.

__C__ **4.** _________ are worn when chipping or grinding.

__D__ **5.** A(n) _________ should always be within easy reach when welding.

A. ground
B. ultraviolet rays
C. safety goggles
D. fire extinguisher
E. overheating

Questions 1-4

Identify the following welding positions.

__C__ **1.** Vertical
__D__ **2.** Horizontal
__A__ **3.** Overhead
__B__ **4.** Flat

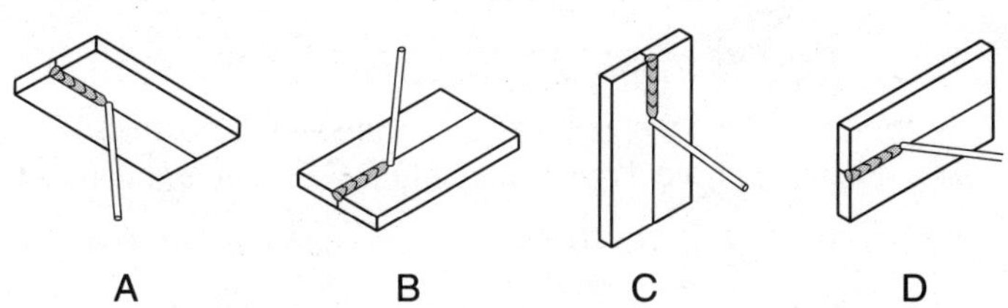

REVIEW QUESTIONS 1

An Essential Skill

True-False

T F 1. Welding is an important means of fabrication used in industry today.

T F 2. Welding is a more expensive process compared to other methods of joining materials.

T F 3. Welding can be used to repair farm, mining, and construction equipment.

T F 4. Oxyacetylene welding uses an intense flame to generate the heat necessary to join metals.

T F 5. A skilled welder must be certified for specific welding jobs.

T F 6. Welding is often used in the construction industry to assemble steel framing in buildings.

T F 7. Welding requires good eyesight and eye-hand coordination.

Multiple Choice

_______ 1. When selecting the proper welding process, which of the following control factors is considered?
  A. kinds of metals to be joined
  B. costs involved
  C. nature of products to be fabricated
  D. production techniques
  E. all of the above

_______ 2. The _______ welding process is used primarily for production welding.
  A. oxyacetylene
  B. shielded metal-arc
  C. gas shielded-arc
  D. resistance
  E. none of the above

_______ 3. The _______ welding process is known for its flexibility and mobility.
  A. oxyacetylene
  B. shielded metal-arc
  C. gas shielded-arc
  D. resistance
  E. none of the above

_______ 4. Forge welding was first used in _______.
  A. 1863
  B. 1475
  C. 2000 B.C.
  D. 1915
  E. none of the above

KNOW YOUR WELDING SYMBOLS

Arrow Side/Other Side of Welding Symbol

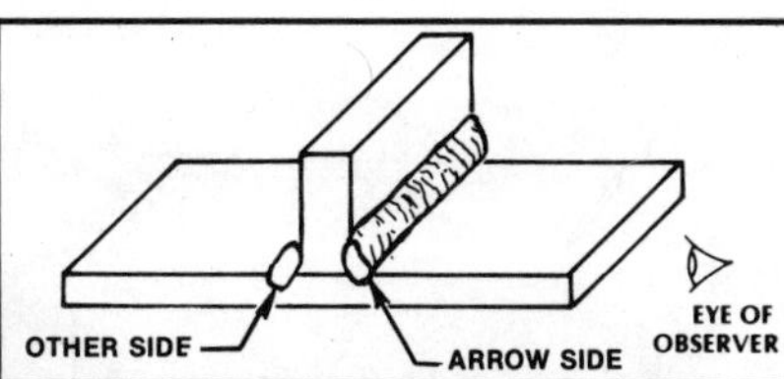

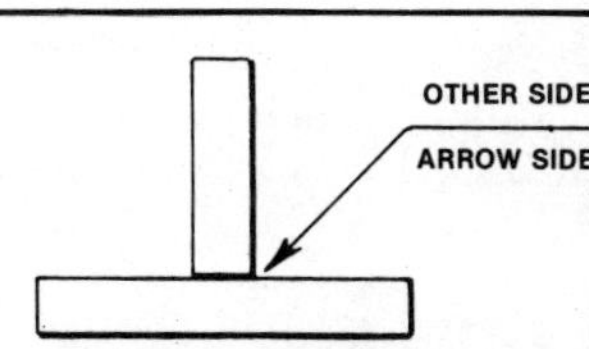

C 5. The gas metal-arc welding process ________.
- A. uses oxygen as a shielding gas
- B. deposits weld metal at a slower rate than shielded metal-arc welding
- C. uses a consumable wire electrode
- D. is no longer used in industry today
- E. all of the above

E 6. A skilled welder ________.
- A. must recognize weld defects
- B. reads weld symbols
- C. has a knowledge of properties of metals
- D. is certified for a specific welding job
- E. all of the above

B 7. Identify the arrow side of the weld joint shown.

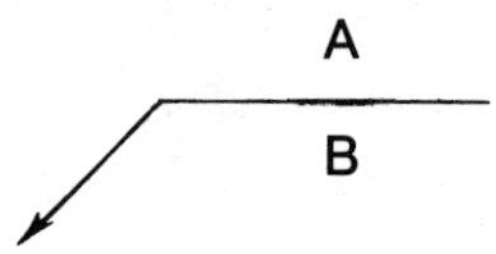

- A. A
- B. B
- C. either
- D. both
- E. none of the above

A 8. Identify the other side of the welding symbol shown.

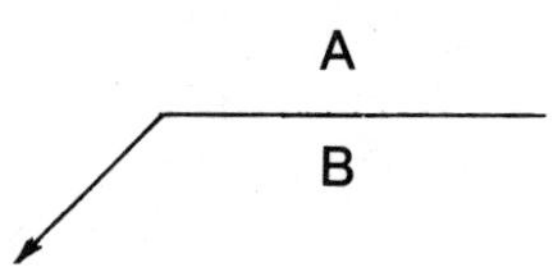

- A. A
- B. B
- C. either
- D. both
- E. none of the above

Matching

B 1. The oxyacetylene welding process ________.

C 2. The shielded metal-arc welding process ________.

A 3. The gas shielded-arc welding process ________.

D 4. The resistance welding process ________.

E 5. Industry today ________.

- A. uses a shielding gas around the arc to prevent atmospheric contamination
- B. uses a mixture of gases to produce an intense flame
- C. deposits molten metal into the joint from the tip of the electrode
- D. uses high current passed through two metal pieces between electrodes
- E. uses several types of welding

REVIEW QUESTIONS 2

Welding Safety

True-False

T F 1. Accidents occur because of carelessness or indifference to safety regulations.

T F 2. Safety must always be practiced.

T F 3. An accident must always be reported no matter how slight it may be.

T F 4. Any form of horseplay in a shop is dangerous and can lead to an accident.

T F 5. Welding equipment should never be used until exact instructions on its operation are received.

T F 6. A malfunction in welding equipment should never be remedied without consulting supervising personnel.

T F 7. Some welding processes do not require a well ventilated area.

T F 8. A respirator should be used in situations when welding metals which give off toxic fumes.

T F 9. Welding or cutting a container that has held combustible material must be cleaned properly to avoid the possibility of an explosion.

Multiple Choice

_______ 1. _________ is not a basic rule that contributes to the safe handling of oxyacetylene equipment.
- A. Opening cylinder valves slowly
- B. Keeping heat, flame, and sparks away from combustibles
- C. Avoiding oxygen and acetylene leaks
- D. Protecting the equipment from rust with a coating of oil
- E. all of the above

_______ 2. When welding and cutting a container, _________.
- A. determine what substance was held in it before cleaning it
- B. never use oxygen to ventilate it
- C. a vent or opening must be provided to release air pressure or steam
- D. never rely on your nose or eyes to determine if it is safe to weld or cut a closed container
- E. all of the above

_______ 3. Which of the following safety practices is not common to all arc welding operations?
- A. Avoid shutting off the power when making repairs to the welding machine to save electricity.
- B. Be sure the welding machine is equipped with a conveniently located power disconnect switch.
- C. Do not weld with loose cable connections.
- D. The polarity switch should not be changed when the machine is under load.
- E. none of the above

KNOW YOUR WELDING SYMBOLS

Fillet Weld Arrow Side

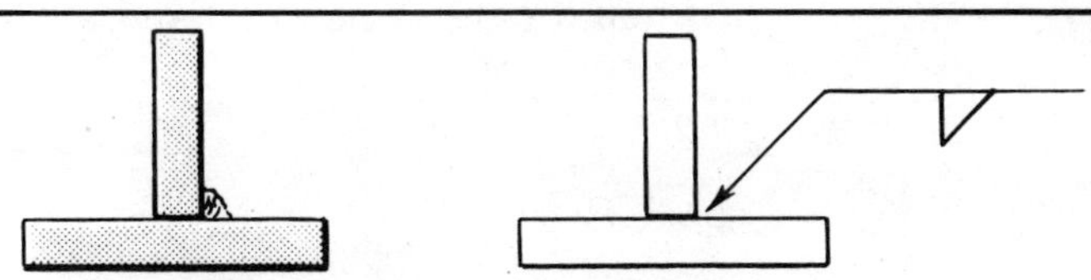

Desired Weld — Symbol

______ **4.** Which of the following is a reason for proper body protection when welding or cutting?
A. Sparks may lead to serious burns.
B. Rays given off are extremely dangerous to the eyes.
C. Hot metal may cause burns to unprotected flesh.
D. Falling slag can travel far and quickly.
E. all of the above

______ **5.** ______ gives off toxic fumes when welding.
A. Aluminum
B. Galvanized steel
C. Medium carbon steel
D. High carbon steel
E. none of the above

______ **6.** What is the meaning of the welding symbol shown?

A. Fillet weld arrow side
B. Fillet weld other side
C. Fillet weld both sides
D. Slot weld arrow side
E. none of the above

Matching

______ **1.** ______ can be harmful to eyes and skin.
______ **2.** A welding machine must have a proper ______ to prevent electrical shock.
______ **3.** Operating with currents above the rated cable capacity causes ______.
______ **4.** ______ are worn when chipping or grinding.
______ **5.** A ______ should always be within easy reach when welding.

A. ultraviolet rays
B. ground
C. safety goggles
D. fire extinguisher
E. overheating

The following statements refer to the proper cleaning methods used prior to welding or cutting a container.

______ **6.** Trisodium phosphate is used in the ______.
______ **7.** Low pressure steam and hot soda are used in the ______.
______ **8.** The ______ is used when the substance is known to be readily soluble in water.
______ **9.** The ______ determines the exact solvents used for cleaning.
______ **10.** The ______ is generally used when the container contains insoluble solvents, or when it cannot be mechanically cleaned.

A. hot chemical solution method
B. water method of cleaning
C. chemical cleaning
D. steam method of cleaning
E. manufacturer

REVIEW QUESTIONS 3

Welding Metallurgy

True-False

T F 1. Chemical properties of metals include corrosion, oxidation, and reduction.

T F 2. A temperature indicating crayon can be used to determine the correct preheat and postweld heat treatment temperature.

T F 3. Mechanical properties determine the behavior of metals under applied loads.

T F 4. The grain structure of metal remains the same when heat is applied.

T F 5. When a metal is cold-worked, the ferrite and pearlite grains are made larger.

T F 6. Annealing is a softening process that allows metal to be machined.

T F 7. Normalizing is a process in which the properties of the metal are changed by rapidly cooling the metal after it is heated.

T F 8. Hardening of a metal is accomplished by heating the metal above its critical point then rapidly cooling it in air, water, oil, or brine.

T F 9. Carburizing is a process in which steel is hardened through the absorption of carbon on its surface.

T F 10. Critical temperature is the temperature of transformation of a metal.

Multiple Choice

_____ 1. A(n) _____ holds the metal being welded in a fixed position.
- A. inclusion
- B. two pass
- C. distortion plate
- D. jig
- E. intermittent weld

_____ 2. _____ are used to control expansion and contraction forces on long seams
- A. Chill blocks
- B. Neutral axes
- C. Fit-ups
- D. Bevel angles
- E. Tack welds

_____ 3. _____ is not used to control distortion when welding.
- A. A chill block
- B. Preheating
- C. Porosity
- D. Neutral axis
- E. Postweld heat treatment

KNOW YOUR WELDING SYMBOLS

Fillet Weld Other Side

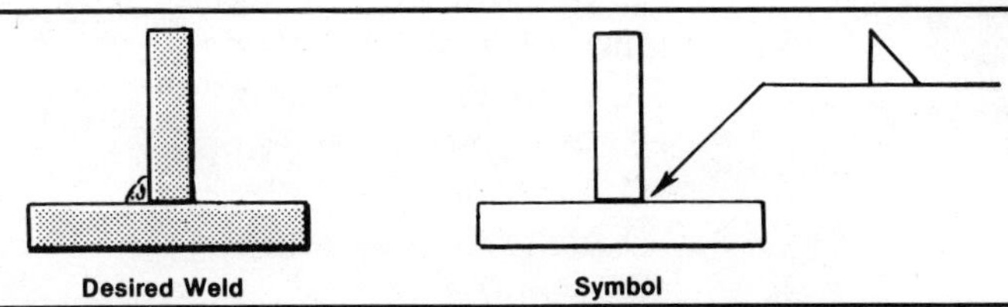

_______ **4.** The physical properties of a metal include _______________.
- A. melting point
- B. thermal conductivity
- C. grain structure
- D. all of the above
- E. none of the above

_______ **5.** _______________ is not a case hardening process.
- A. Annealing
- B. Carburizing
- C. Cyaniding
- D. Nitriding
- E. all of the above

_______ **6.** The ___________ type space-lattice is found in such metals as iron and molybdenum.
- A. body-centered cube
- B. face-centered cube
- C. close packed hexagonal form
- D. all of the above
- E. none of the above

_______ **7.** Metal _________________ when heated.
- A. contracts
- B. expands
- C. remains the same
- D. all of the above
- E. none of the above

_______ **8.** What is the meaning of the welding symbol shown?

- A. Fillet weld arrow side
- B. Fillet weld other side
- C. Fillet weld both sides
- D. Slot weld arrow side
- E. none of the above

_______ **9.** A mixture of pearlite and ferrite with less than .83% carbon is called _______________.
- A. eutectoid
- B. pearlite
- C. hypereutectoid
- D. hypoeutectoid
- E. cold-worked

_______ **10.** Identify the welding technique shown.

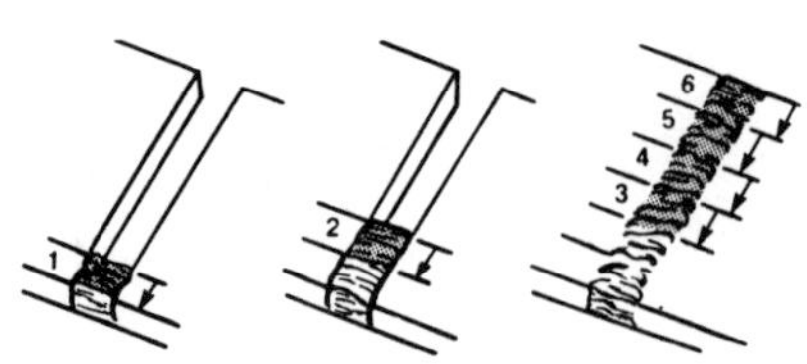

- A. backstep welding
- B. intermittent welding
- C. tack welding
- D. distortion welding
- E. peening

_______ **11.** Identify the welding technique shown.

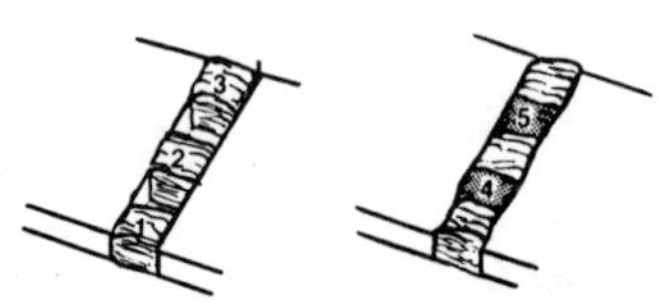

- A. back-step welding
- B. intermittent welding
- C. tack welding
- D. distortion welding
- E. peening

Matching

1. __________ is the internal resistance a material offers.
2. __________ is the deformation that results from stress.
3. __________ is the ability of a metal to return to its original shape.
4. __________ is the last point at which a metal may be stretched and still return to its original condition.
5. __________ is the ratio of stress to strain within the elastic limit.

A. Elasticity
B. Strain
C. Stress
D. Modulus of elasticity
E. Elastic limit

6. __________ is the property of a metal that allows it to resist forces acting to pull the metal apart.
7. __________ is the ability of a metal to resist being crushed.
8. __________ is the quality of a metal that resists forces from causing a member to bend or deflect in the direction of the applied load.
9. __________ is the ability of a metal to withstand forces that cause a member to twist.
10. __________ is the ability of a metal to withstand two equal forces acting in opposite directions.

A. Shear strength
B. Torsional strength
C. Tensile strength
D. Bending strength
E. Compressive strength

11. __________ of a metal is its ability to resist various kinds of rapidly alternating stresses.
12. __________ is the ability of a metal to resist loads that are applied suddenly.
13. __________ is the ability of a metal to stretch, bend, or twist without breaking or cracking.
14. __________ is the property of a metal that resists indentation or penetration.
15. __________ is the property of a metal that will fracture under stress because of low ductility.

A. Ductility
B. Impact strength
C. Brittleness
D. Hardness
E. Fatigue strength

16. The formation of tiny pinholes caused by atmospheric contamination is called __________.
17. Heating the structure before welding is called __________.
18. Heating the structure after welding takes place is called __________.
19. Hammering the weld with the round end of a ball peen hammer after welding is called __________.
20. The heat treating process in which the welded component is placed in a furnace is called __________.

A. peening
B. postweld heat treatment
C. preheating
D. stress relieving
E. porosity

______ 21. ______ is an alloying element which when added to steel results in a metal with greater hardness and resistance to wear without becoming brittle.		
______ 22. ______ added to steel increases its ductility while maintaining its strength.		A. Nickel
______ 23. ______ results from the wide range between the temperature of the base metal and the molten weld metal.		B. Blowholes
______ 24. ______ occurs from gas entrapment during solidification of the weld metal.		C. Chromium
______ 25. ______ is caused by impurities forced into the molten puddle during the welding process.		D. Grain growth
		E. Inclusions

______ 21. ______ is an alloying element which when added to steel results in a metal with greater hardness and resistance to wear without becoming brittle.

______ 22. ______ added to steel increases its ductility while maintaining its strength.

______ 23. ______ results from the wide range between the temperature of the base metal and the molten weld metal.

______ 24. ______ occurs from gas entrapment during solidification of the weld metal.

______ 25. ______ is caused by impurities forced into the molten puddle during the welding process.

A. Nickel
B. Blowholes
C. Chromium
D. Grain growth
E. Inclusions

______ 26. ______ is the property of a metal that has strength combined with ductility.

______ 27. ______ is the ability of a metal to be deformed by compression forces without developing defects.

______ 28. ______ is a slow but progressively increasing strain at high temperatures, causing the metal to fail.

______ 29. ______ is the behavior characteristics of metals when in a low temperature environment.

______ 30. ______ refers to the amount of expansion in one inch or foot produced by a temperature rise of 1°F.

A. Creep
B. Coefficient of expansion
C. Toughness
D. Cryogenic property
E. Malleability

______ 31. Low carbon steels ______.

______ 32. Medium carbon steel ______.

______ 33. High carbon steels ______.

______ 34. Very-high carbon steels ______.

A. contain 0.75%–1.7% carbon
B. throw long white streamers with little or no sparklers when spark tested.
C. contain 0.45%–0.75% carbon
D. contain 0.30%–0.45% carbon

Questions 35–39. Steel Code Classifying Systems

Identify the parts of the alloy steel code shown:

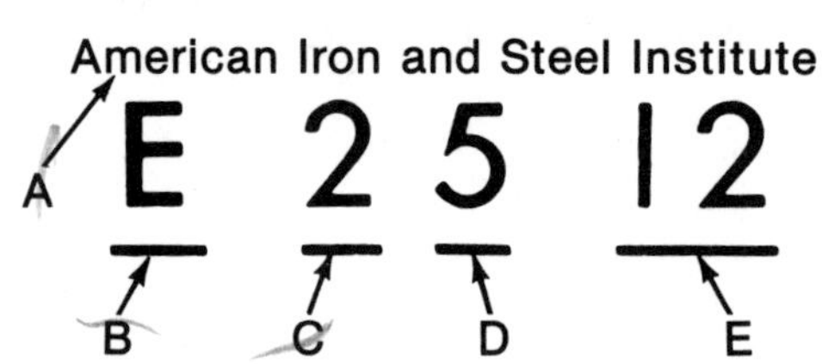

______ 35. amount of nickel

______ 36. electric furnace

______ 37. amount of carbon

______ 38. nickel steel

______ 39. AISI

REVIEW QUESTIONS 4

Joint Design and Welding Terms

True-False

T F 1. The four main welding positions are flat, horizontal, vertical, and overhead.

T F 2. Weld joint design is greatly influenced by the cost of preparing the joint.

T F 3. The correct groove bevel commonly used has an included angle of 60°.

T F 4. A fillet is one of the five basic weld joints.

T F 5. A square butt joint does not require edge preparation.

T F 6. A double fillet lap joint is welded on one side.

T F 7. A single V-butt joint is used on plate $^3/_8$″ or greater in thickness.

T F 8. For joint efficiency in a lap joint, an overlap greater than three times the thickness of the thinnest member is recommended.

T F 9. Overhead position welding is easier to learn than flat position welding.

T F 10. Fit-up must be consistent for the entire length of the joint to obtain a proper weld.

Matching

______ 1. The ______ is the metal to be welded.

______ 2. The narrow layer of weld metal deposited is call a ______.

______ 3. ______ is the shape within the deposited bead caused by the welding heat.

______ 4. Each additional bead deposited on the base metal is called a ______.

______ 5. The ______ is the depression in the base metal made by the welding heat source.

A. pass
B. base (parent) metal
C. ripple
D. bead
E. crater

______ 6. The ______ is the depth of fusion with the base metal.

______ 7. The ______ is the amount of weld metal deposited above the surface of the pieces being joined.

______ 8. The bottom lip near the slanted surface of the groove is called the ______.

______ 9. The distance from toe to toe across the face of the weld is the ______.

______ 10. The point where the weld metal meets the base metal is called the ______.

A. root face
B. penetration
C. toe
D. weld width
E. reinforcement

KNOW YOUR WELDING SYMBOLS

Fillet Weld Both Sides

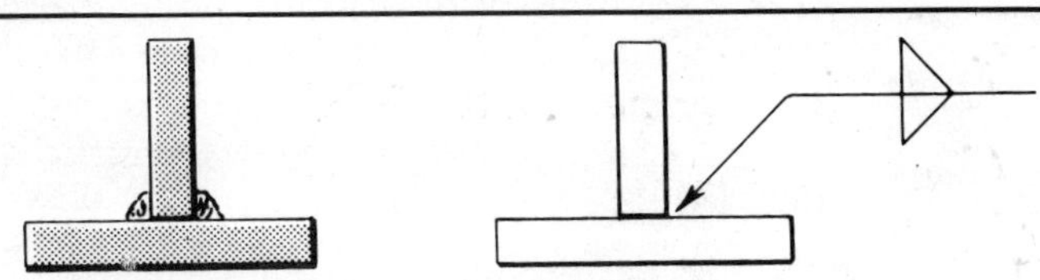

Desired Weld   Symbol

______ **11.** The exposed surface of the weld bounded by the toes of the weld is called the ______ of the weld.

______ **12.** The ______ is the point of the weld triangle opposite the face of the weld.

______ **13.** The distance through the center of the weld from the face to the root is called the ______.

______ **14.** The size of fillet welds made in lap or T-joints is determined by the ______.

______ **15.** The ______ is the distance between the root faces of opposite pieces in a groove weld.

A. throat
B. face
C. root
D. root opening
E. weld legs

Questions 16–20
Identify the weld joints shown.

______ **16.** T-joint

______ **17.** Lap joint

______ **18.** Edge joint

______ **19.** Butt joint

______ **20.** Corner

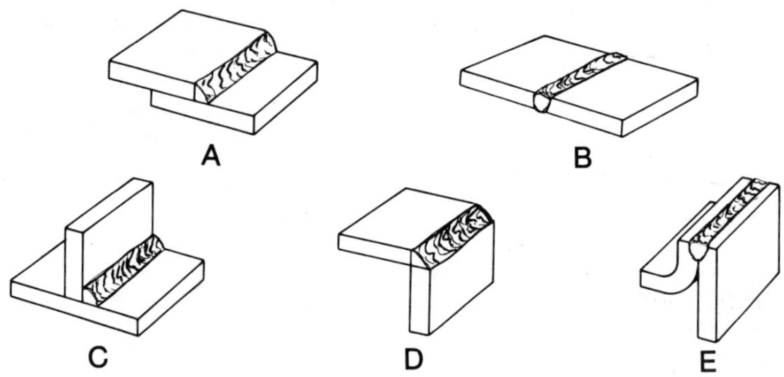

Questions 21–25
Identify the weld types shown.

______ **21.** Surfacing

______ **22.** Fillet

______ **23.** Groove

______ **24.** Plug

______ **25.** Slot

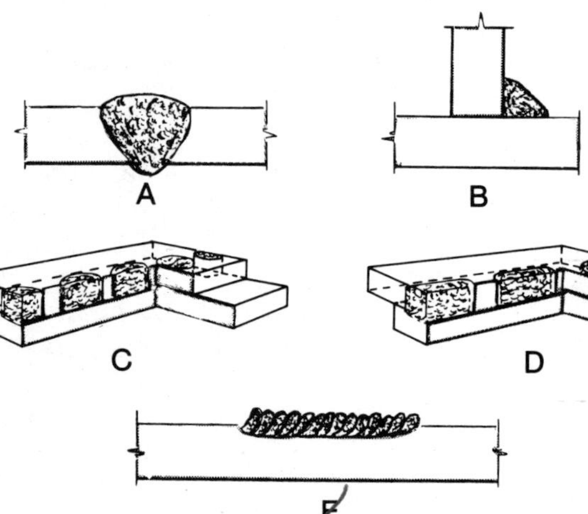

Questions 26–30
Identify the types of butt joint designs shown.

______ **26.** Square butt joint

______ **27.** Single bevel butt joint

______ **28.** Single U-butt

______ **29.** Double V-butt joint

______ **30.** Single V-butt joint

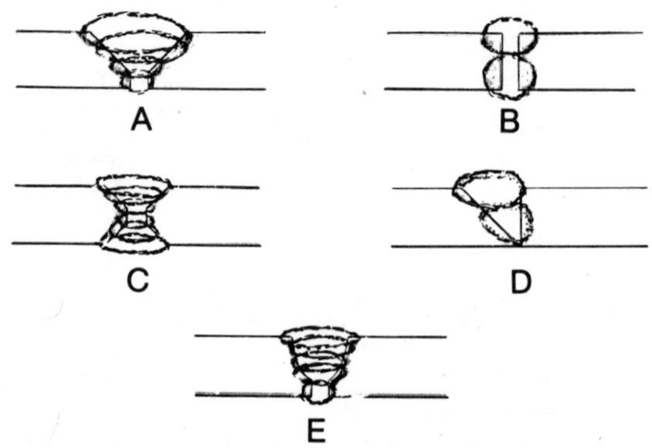

Questions 31–35
Identify the types of T-joint designs shown.

______ **31.** Square T-joint

______ **32.** Single bevel T-joint

______ **33.** Double bevel T-joint

______ **34.** Single-J T-joint

______ **35.** Double-J T-joint

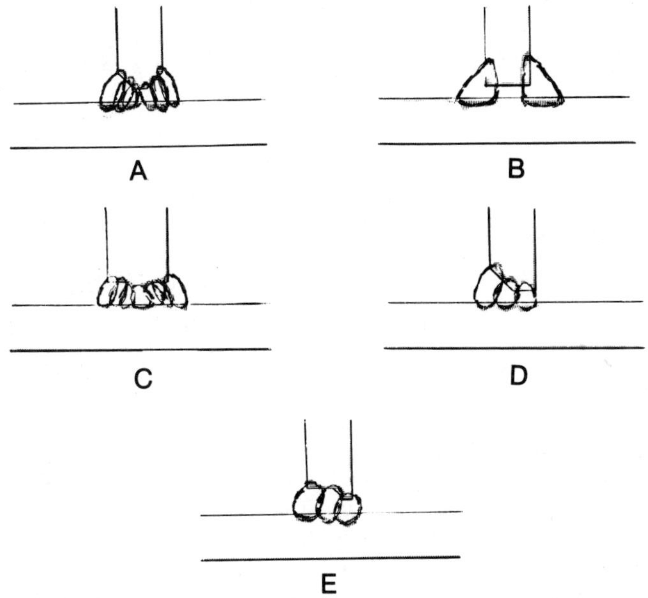

Questions 36–40
Identify the weld joint designs shown.

______ **36.** Double fillet lap joint

______ **37.** Flush corner joint

______ **38.** Half-open corner joint

______ **39.** Full-open corner joint

______ **40.** Edge joint

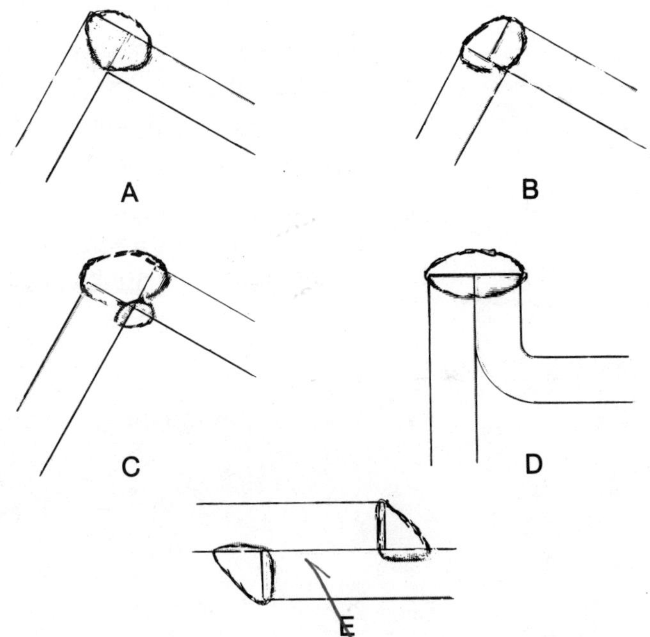

REVIEW QUESTIONS 5

Equipment

True-False

T F 1. The oxyacetylene flame reaches a temperature of 5,700°–6,300°F.

T F 2. The oxygen cylinder valve is opened with a wrench.

T F 3. Acetylene becomes unstable if it is compressed to more than 15 pounds per square inch.

T F 4. Acetylene has no detectable odor.

T F 5. When moving an oxygen or acetylene cylinder, the cylinder should be carefully rolled on its bottom edge.

T F 6. Two types of oxyacetylene welding torches are the injector and the equal pressure.

T F 7. Welding tip size is governed by the diameter of the opening.

T F 8. A pair of pliers is used to mount the tip on the welding torch.

T F 9. A sparklighter should be used to light the oxyacetylene torch.

T F 10. Mapp gas is more sensitive to shock than acetylene.

T F 11. Needle valves on the torch regulate the flow of oxygen and acetylene.

T F 12. Two types of regulators are the single-stage and the two-stage.

T F 13. Welding hoses can be safely repaired with tape.

T F 14. Oxygen and acetylene cylinders should be stored in a vertical position and chained against a stationary object.

T F 15. Cylinders should not be exposed to any open fire, heat, or sparks from the torch.

Multiple Choice

________ 1. What is the meaning of the welding symbol shown?

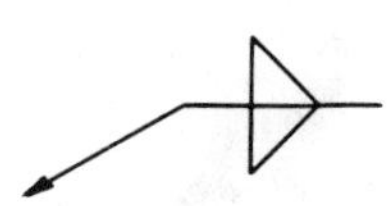

A. Fillet weld arrow side
B. Fillet weld other side
C. Fillet weld both sides
D. Slot weld arrow side
E. none of the above

KNOW YOUR WELDING SYMBOLS

Square-Groove Weld

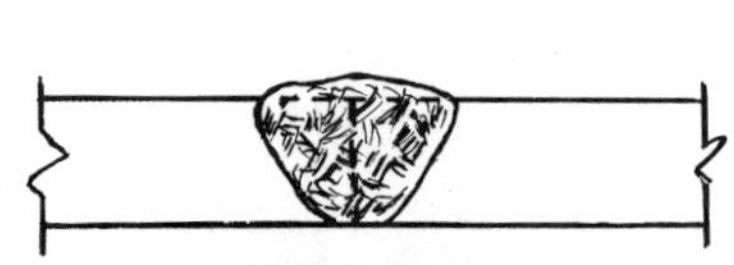

Desired Weld

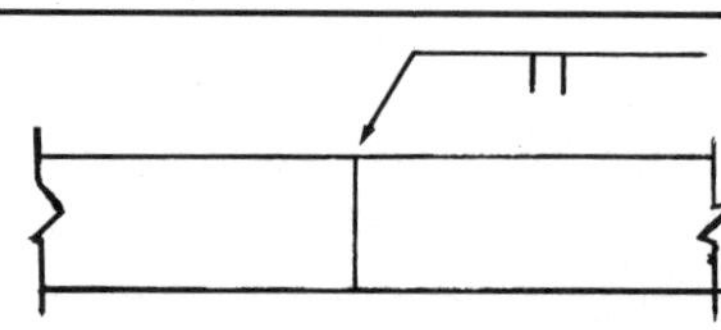

Symbol

______ **2.** Which of the following is not true about acetylene gas?
- A. It becomes very unstable if compressed to more than 15 psi.
- B. It has a very distinctive nauseating odor.
- C. The cylinder valve must be opened all the way when it is in use.
- D. It is colorless.
- E. none of the above

______ **3.** A(n) ______ is necessary for a complete set of welding apparatus.
- A. oxygen regulator
- B. acetylene regulator
- C. welding torch
- D. striker
- E. all of the above

______ **4.** ______ can be used to determine the correct connections to be made when using oxygen and acetylene gases.
- A. Left-hand threads
- B. Right-hand threads
- C. Notched connecting nuts
- D. Hose color
- E. all of the above

______ **5.** Which of the following rules pertaining to the care of regulators is not true?
- A. If the regulator does not function properly, turn off the gas supply and have qualified personnel examine it.
- B. Do not interchange oxygen and acetylene regulators.
- C. Always keep a tight connection between the regulator and the cylinder.
- D. Oil or grease the regulator to ensure smooth operation.
- E. all of the above

______ **6.** What is the meaning of the welding symbol shown?

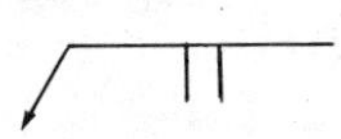

- A. Fillet weld arrow side
- B. Slot weld other side
- C. Square-groove weld arrow side
- D. Square-groove weld other side
- E. none of the above

Matching

______ **1.** In the acetylene cylinders, ______ absorbs large quantities of acetylene.

______ **2.** A ______ prevents damage to the oxygen valve.

______ **3.** The ______ cylinder is charged at a pressure of 2,200 psi.

______ **4.** A ______ prevents flashback from reaching the acetylene manifold system.

______ **5.** ______ becomes unstable if compressed to more than 15 psi.

- A. flash arrestor
- B. oxygen
- C. acetylene
- D. protector cap
- E. acetone

______ **6.** A ______ is used to clean a welding tip.

______ **7.** A ______ transports oxygen.

______ **8.** A ______ transports acetylene gas.

______ **9.** A ______ is found on an oxygen hose.

______ **10.** A ______ is found on an acetylene hose.

- A. red hose
- B. green hose
- C. tip cleaner
- D. right-hand thread
- E. left-hand thread

REVIEW QUESTIONS 6

Setting-Up and Operating

True-False

T F 1. Applying soapy water to an oxyacetylene outfit with a brush is the best method to locate leaks.

T F 2. The cylinder valve is cracked to blow out dirt lodged in the outlet nozzle.

T F 3. A close-fitting wrench is used to attach regulators to cylinders.

T F 4. Welding hoses are purged by opening the cylinder valve.

T F 5. The size of a welding tip depends on the amount of pressure used.

T F 6. The welder should stand to one side of the regulator when opening the cylinder valves.

T F 7. The acetylene needle valve should be opened fully to ensure quick lighting of the torch.

T F 8. A torch should be lit with the tip pointing downward.

T F 9. Cylinders should be secured with chains to prevent accidental tipping.

T F 10. If a striker is unavailable, a cigarette lighter may be used to light the torch.

Multiple Choice

__________ 1. The __________ must be tested for leaks.
- A. oxygen and acetylene needle valves
- B. oxygen cylinder valve
- C. acetylene cylinder valve
- D. oxygen regulator inlet connection
- E. all of the above

__________ 2. Which of the following is *not* a correct procedure to follow when lighting the torch?
- A. Stand to one side when opening the cylinder valve.
- B. Open the oxygen cylinder valve all the way.
- C. Open the acetylene cylinder valve all the way.
- D. Use a striker to light the welding torch.
- E. Adjust the gas flow with the needle valves to obtain a neutral flame.

__________ 3. __________ is the last step to take when shutting off the entire welding unit after all welding is complete.
- A. Igniting the flame to burn residual gases out of the hoses
- B. Shuting off both the acetylene and oxygen cylinder valves
- C. Removing pressure on the working gauges by opening the needle valves until the lines are drained, then promptly close the needle valves
- D. Releasing the adjusting screws on the pressure regulators
- E. none of the above

KNOW YOUR WELDING SYMBOLS

Single-V-Groove Weld

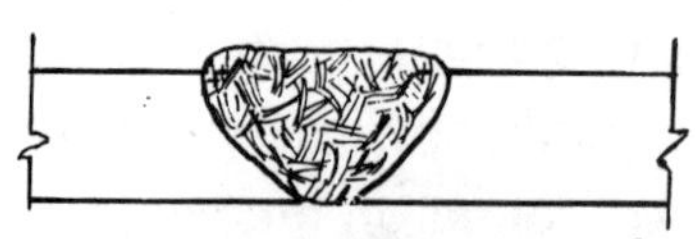

Desired Weld

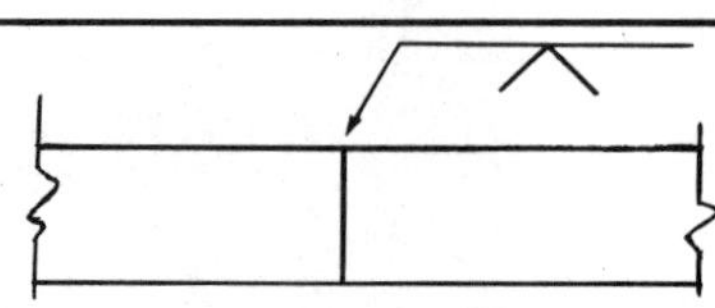

Symbol

_____ **4.** What is the meaning of the welding symbol shown?

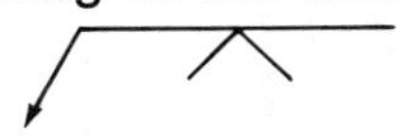

A. Groove weld arrow side
B. Single V-groove weld arrow side
C. Fillet weld arrow side
D. Butt weld other side
E. Fillet weld other side

Matching

_____ **1.** The welder must _____ the welding hoses to remove all residual gases.
_____ **2.** The _____ is determined by the thickness of the metal being welded.
_____ **3.** The _____ should not touch the pieces being welded.
_____ **4.** The _____ is submerged in water to check for leaks.
_____ **5.** The _____ prevents the reverse flow of gases.

A. welding hose
B. check valves
C. purge
D. welding torch tip
E. torch tip number

_____ **6.** A(n) _____ flame results from a one-to-one mixture of gases.
_____ **7.** A(n) _____ flame has a slight excess of acetylene.
_____ **8.** A welding torch marked _____ indicates oxygen.
_____ **9.** A(n) _____ flame has a slight excess of oxygen.
_____ **10.** A welding torch marked _____ indicates acetylene.

A. AC
B. OX
C. neutral
D. oxidizing
E. carburizing

_____ **11.** An acetylene feather is present beyond the inner cone of the _____ flame.
_____ **12.** A(n) _____ occurs when the flame goes out with a loud pop.
_____ **13.** The inner cone of the _____ flame shortens, and the flame burns with a decided roar.
_____ **14.** A flame burning inside the torch with a shrill hissing noise is called a(n) _____.
_____ **15.** A _____ noise made by the torch indicates that there is an insufficient amount of gas flowing to the tip.

A. backfire
B. flashback
C. popping
D. carburizing
E. oxidizing

_____ **16.** Characteristics of the torch flame are controlled by _____ at the torch.
_____ **17.** Acetylene burned at a low pressure without the addition of oxygen produces _____.
_____ **18.** The first gas ignited when preparing to weld is _____.
_____ **19.** The first gas shut off when welding is complete is _____.
_____ **20.** Gas flow to the regulators is controlled by _____.

A. acetylene
B. oxygen
C. cylinder valves
D. needle valves
E. smoke

REVIEW QUESTIONS 7

Flat Position

True-False

T F 1. A welding torch is held at a 45° angle to the work when welding in the flat position.

T F 2. When carrying a puddle without a filler rod, the inner cone of the flame is held approximately 1/8" from the work.

T F 3. A filler rod with a diameter approximately equal to the thickness of the metal being welded should be used.

T F 4. When welding in the flat position, the filler rod is held at a 90° angle.

T F 5. Properties of the metal being welded determine the type of filler rod to be used.

T F 6. A filler rod will stick to metal that is too hot.

T F 7. Tack welds are used to maintain consistent spacing on the weld joint before the weld bead is deposited.

T F 8. The welding torch should be manipulated when welding a butt joint.

T F 9. The heat of a torch should be concentrated on the base metal when running beads with a filler rod.

T F 10. The filler rod should be positioned in the welding puddle during the entire length of the joint.

T F 11. A welding torch may be gripped the way a hammer or a pencil is gripped.

T F 12. Filler rods range in diameters from 1/16" to 3/8".

T F 13. The filler rod should be dipped into the center of the molten pool.

T F 14. A semicircular or circular torch movement should be used when running stringer beads.

T F 15. A flange joint may be welded without adding filler metal.

Multiple Choice

________ 1. Which of the following statements is true when using a filler rod for oxyacetylene welding?

A. The rod should be dipped into the middle of the molten pool when using a circular torch motion.
B. If the filler rod diameter is too large, the heat of the pool will be insufficient to melt the rod.
C. If the rod diameter is too small, the heat cannot be absorbed by the rod, resulting in a hole in the welded plate.
D. The rod should not be held above the molten pool.
E. all of the above

KNOW YOUR WELDING SYMBOLS

Double-V-Groove Weld

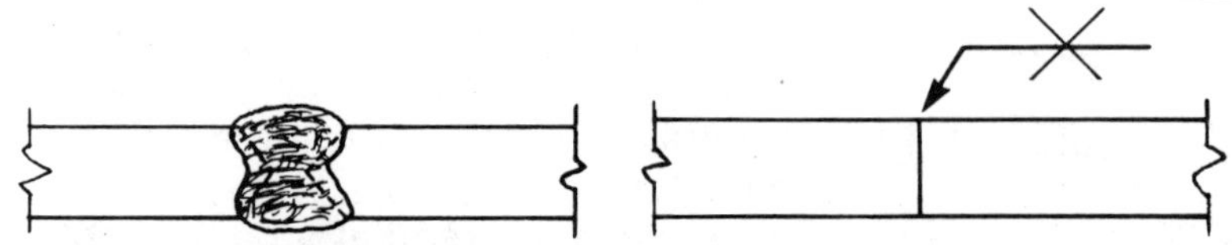

Desired Weld — Symbol

_______ 2. Which of the following is not a recommended procedure for welding a lap joint?
A. Use a semicircular motion with the torch.
B. Direct more heat to the top plate.
C. Increase the duration of the torch motion on the bottom plate.
D. Dip the filler metal into the middle of the molten puddle.
E. all of the above

_______ 3. The _______ of a metal determines the diameter of the filler rod used.
A. composition
B. thickness
C. temperature
D. surface
E. all of the above

_______ 4. When welding a flange joint or a corner joint, _______ is not required for a strong weld.
A. tack welding
B. preheating
C. filler rod
D. a neutral flame
E. all of the above

_______ 5. What is the meaning of the welding symbol shown?

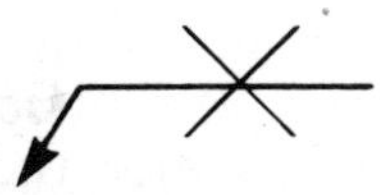

A. Double V-groove weld
B. V-groove weld arrow side
C. Double fillet
D. Double butt
E. none of the above

Matching

_______ 1. Poor penetration across a bead is caused by _______.
_______ 2. A brittle weld is caused by _______.
_______ 3. Holes in joints are caused by _______.
_______ 4. Excessive penetration is caused by _______.
_______ 5. A hole in the end of a joint is caused by _______.

A. a travel speed too slow
B. a travel speed too fast
C. an improper flame adjustment
D. holding a flame too long in one spot
E. not lifting the torch at the end of the joint

REVIEW QUESTIONS 8

Other Welding Positions

True-False

T F **1.** Welding in a flat position is easier and faster than welding in a horizontal, vertical, or overhead position.

T F **2.** The amount of heat directed to the weld joint in the overhead position should be minimized to prevent the weld puddle from getting too large.

T F **3.** When practicing welds in positions other than the flat, a jig or positioner can be used.

T F **4.** When welding a butt joint in the horizontal position, the flame should be directed above the joint.

T F **5.** Welding a T-joint in a horizontal position requires that the torch be held at a 45° angle to the flat plate and to the line of weld.

T F **6.** When welding in the overhead position, precautions must be taken to prevent hot molten metal from causing injury.

T F **7.** When welding a butt joint in the vertical position, more of the flame should be directed on the weld puddle.

T F **8.** Welding in the overhead position requires semicircular torch manipulation.

T F **9.** If the puddle becomes too large, the flame must be raised slightly away from the work.

T F **10.** Overhead welding is possible because of the cohesive characteristics of molten metal.

Multiple Choice

__________ **1.** The main obstacle that prevents a sound weld in positions other than flat is the ____________ downward on the molten metal.

A. filler rod motion
B. heat expansion
C. cold lap
D. gravitational pull
E. all of the above

__________ **2.** Filler rod is added ____________ to prevent undercutting when welding a T-joint in the horizontal position.

A. closer to the horizontal plate
B. close to the inner cone of the welding torch
C. on top of the torch flame
D. closer to the vertical plate
E. none of the above

KNOW YOUR WELDING SYMBOLS

Single-Bevel-Groove Weld

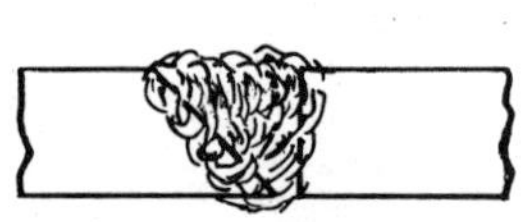

Desired Weld

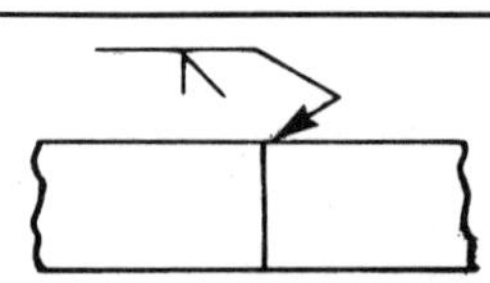

Symbol

_______ **3.** More of the flame is directed on the _______ to prevent the puddle from becoming too fluid when welding a butt joint in the vertical position.
   A. base metal
   B. filler rod
   C. right plate
   D. left plate
   E. none of the above

_______ **4.** When welding a T-joint in horizontal position, the tip of the torch is held at a _______° angle to the flat plate and to the line of the weld.
   A. 15
   B. 20
   C. 30
   D. 45
   E. 90

_______ **5.** When welding a butt joint in an overhead position, the _______ to prevent the puddle from running.
   A. flame is increased
   B. angle of the filler rod is changed
   C. torch is pulled slightly away
   D. torch is positioned closer to the work
   E. all of the above

_______ **6.** What is the meaning of the welding symbol shown?

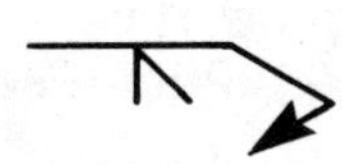

   A. Bevel-groove arrow side
   B. Groove other side
   C. Bevel-groove other side
   D. Groove arrow side
   E. none of the above

REVIEW QUESTIONS 9

Heavy Steel Plate

True-False

T F **1.** In industry, the oxyacetylene process is frequently used to weld heavy steel plate.

T F **2.** Steel that is less than ¼″ thick is generally called sheet.

T F **3.** The oxyacetylene welding techniques used for heavy steel plates are the same as those used for lighter gauges of steel.

T F **4.** When welding heavy steel plate, edge preparation is necessary for proper penetration.

T F **5.** The edges of a metal 1/8″ thick or less must be beveled.

T F **6.** The recommended included angle for single V-butt joints is 60°.

T F **7.** Filler rods that have similar composition to the base metal should be used when welding high-carbon steel.

T F **8.** The backhand welding technique is used for welding lighter gauges of steel.

T F **9.** High-carbon steels that have been welded must be heat treated to regain their original properties.

T F **10.** Plates thicker than ½″ require several passes to fill the V of the weld joint.

Multiple Choice

________ **1.** Metal thicker than __________″ is designated as plate.
- A. 1/16
- B. 1/8
- C. 3/16
- D. 1/4
- E. 1/2

________ **2.** Using the __________ technique, the filler rod follows the welding torch in the direction of the weld.
- A. overhand
- B. backhand
- C. forehand
- D. weave
- E. none of the above

________ **3.** Using the __________ technique, the filler rod comes before the welding torch in the direction of the weld.
- A. overhand
- B. backhand
- C. underhand
- D. weave
- E. none of the above

KNOW YOUR WELDING SYMBOLS

Single-Bevel-Groove Weld

Desired Weld

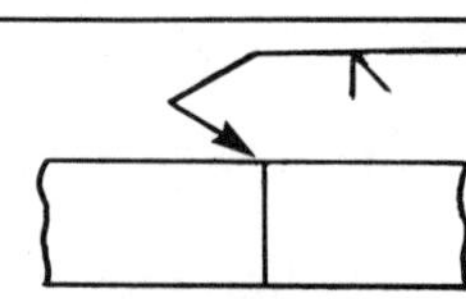

Symbol

_______ **4.** When welding a single V-butt joint on heavy steel plate, the torch should be held with the flame at a _______° angle from vertical.
A. 30
B. 45
C. 60
D. 75
E. 90

_______ **5.** Grind marks on a specimen prepared for testing should run _______ on the piece to prevent premature weld failure.
A. circular
B. lengthwise
C. widthwise
D. at a 45° angle
E. none of the above

_______ **6.** A _______ joint is recommended for plates up to ½″ thick.
A. double V
B. square butt
C. single V
D. closed butt
E. all of the above

_______ **7.** Steels containing _______ % of carbon or more are considered high-carbon steels.
A. 0.21
B. 0.37
C. 0.67
D. 0.89
E. none of the above

_______ **8.** Heavy steel plate butt welds made with the oxyacetylene process should be tested with a _______.
A. hammer and chisel
B. guided bend tester
C. pair of pliers
D. compression tester
E. all of the above

_______ **9.** The _______ flame should be used when welding high-carbon steel plate using the oxyacetylene process.
A. oxidizing (excess oxygen)
B. slightly carburizing (excess acetylene)
C. neutral
D. harsh
E. none of the above

_______ **10.** What is the meaning of the welding symbol shown?

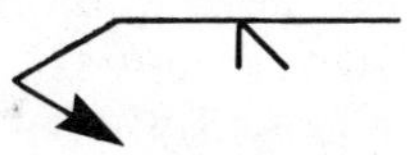

A. Bevel-groove weld arrow side
B. Bevel-groove weld other side
C. Groove weld
D. Fillet weld other side
E. none of the above

REVIEW QUESTIONS 10

Gray Cast Iron

True-False

T F 1. Preheating and postweld heat treatment are necessary when welding gray cast iron.

T F 2. If a casting is cooled too rapidly, the weld area is likely to turn into white cast iron.

T F 3. The edges of the casting should be beveled to have a 45° included angle.

T F 4. The weld area should be cleaned at least 1″ on both sides of the V.

T F 5. A correct filler rod must be used to maintain the properties of the casting through the weld area.

T F 6. A large casting is placed in a furnace to ensure proper preheating.

T F 7. When preheating, cast iron should be heated to a bright red.

T F 8. A cast iron filler rod contains silicon to assure soft, machinable weld deposits when used with the correct preheat and postweld heat treatment.

T F 9. The inner cone of the flame should be held approximately 1/8″ to 1/4″ from the joint.

T F 10. The filler rod should be dipped in and out of the weld puddle as welding proceeds.

T F 11. Gray cast iron is extremely brittle and is susceptible to temperature changes.

T F 12. The filler rod can be used to skim off impurities in the weld puddle.

Multiple Choice

________ 1. Flux is essential when welding gray cast iron because it ________.
- A. keeps the molten puddle fluid
- B. prevents iron oxides from forming
- C. prevents inclusions and blowholes from forming
- D. all of the above
- E. none of the above

________ 2. Which of the following is not a correct procedure to follow when welding gray cast iron?
- A. Move the flame from side to side of the V.
- B. Keep the torch in the same position as in welding mild steel.
- C. Insert the fluxed end of the filler rod into the molten pool.
- D. Hold the torch in the same location to remove gas bubbles in the puddle.
- E. none of the above

________ 3. Preheating a casting before welding ________.
- A. can be done in a furnace
- B. minimizes the possibility of cracks developing
- C. equalizes the expansion and contraction forces
- D. requires heating the entire casting to a dull red
- E. all of the above

KNOW YOUR WELDING SYMBOLS

Double-Bevel-Groove Weld

Desired Weld

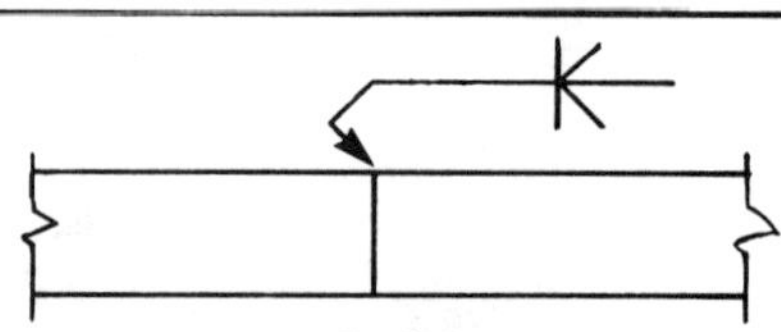

Symbol

_______ **4.** What is the meaning of the welding symbol shown?

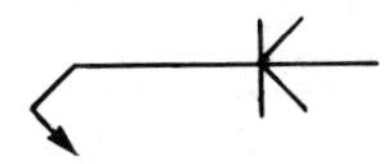

A. Fillet weld both sides
B. Butt joint both sides
C. Groove both sides
D. Bevel arrow side
E. none of the above

Matching

_______ **1.** Correct _______ is applied to the filler rod.

_______ **2.** A casting receives _______ after the welding is complete.

_______ **3.** Improperly cleaned surfaces cause _______.

_______ **4.** The _______ should have the same composition as the base metal.

_______ **5.** Playing the flame over the casting is a common method of _______.

A. preheating
B. postweld heat treatment
C. flux
D. filler rod
E. blowholes

REVIEW QUESTIONS 11

Aluminum

True-False

T F **1.** Aluminum is very weak when heated and care must be taken to adequately support it during the welding operation.

T F **2.** The thermal conductivity of aluminum is approximately four times that of steel.

T F **3.** Aluminum turns dull red when heated to the melting point.

T F **4.** The proper flux is required to prevent oxides from forming during the welding process.

T F **5.** Aluminum has a melting point of 2,200°F.

T F **6.** Lap welds are not recommended when oxacetylene welding aluminum.

T F **7.** If possible, aluminum welds should be made in single pass.

T F **8.** The weld area should be thoroughly cleaned before welding.

T F **9.** All aluminum to be welded is preheated to prevent distortion from expansion and to minimize cracks.

T F **10.** The edges of $^{1}/_{16}''$ to $^{3}/_{16}''$ plate are notched to allow complete penetration of the joint.

Multiple Choice

__________ **1.** Which of the following is a recommended procedure for oxyacetylene welding of aluminum?

A. A neutral or reducing flame is used.
B. The work is preheated before welding.
C. Flux must be used to ensure a sound weld.
D. Impurities on the metal must be cleaned before welding.
E. all of the above

__________ **2.** ______________ is not used to determine the correct preheat temperature of a metal.

A. Striking the metal with a hammer
B. A pine stick
C. Carpenter's chalk
D. A temperature indicating crayon
E. Noting color change in the metal

__________ **3.** ______________ is not a correct procedure to follow when welding aluminum using the oxyactylene process.

A. Using a forehand technique
B. Holding the filler rod in the puddle through the entire weld
C. Angling the torch at a low angle
D. Passing the flame over the starting point until the flux melts
E. none of the above

KNOW YOUR WELDING SYMBOLS

J-Groove Weld

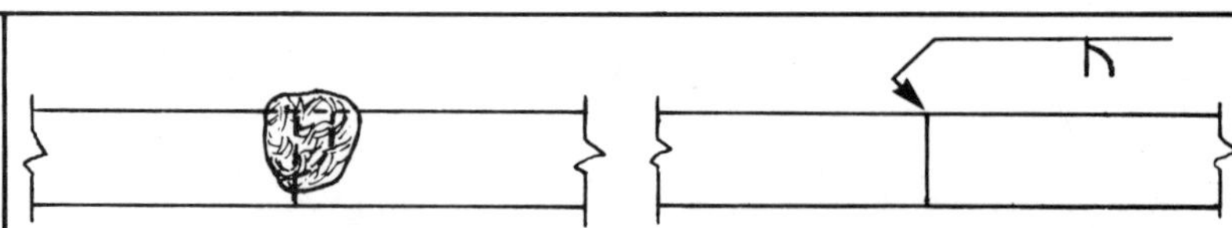

Desired Weld Symbol

________ **4.** ________ filler rods are commonly used for nonheat-treatable aluminum.
- A. 1100
- B. 4043
- C. 5356
- D. all of the above
- E. none of the above

________ **5.** A filler rod ________" in size is commonly used when welding aluminum using the oxyacetylene process.
- A. 1/16
- B. 1/8
- C. 3/16
- D. 1/4
- E. all of the above

________ **6.** What is the meaning of the welding symbol shown?

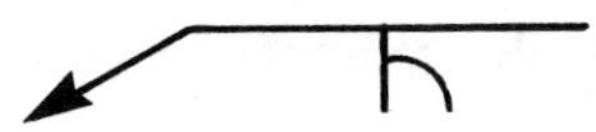

- A. J-groove weld other side
- B. J-groove weld arrow side
- C. J-groove weld both sides
- D. V-groove weld other side
- E. V-groove weld arrow side

Matching

________ **1.** A type of filler rod commonly used for welding aluminum is ________.

________ **2.** The torch angle used when welding thin aluminum is ________°.

________ **3.** The melting point of aluminum is ________°F.

________ **4.** The recommended preheat temperature for aluminum plate ¼" or more in thickness is ________°F.

________ **5.** The included angle for butt joints 3/16" or more in thickness is ________°.

- A. 30
- B. 90–120
- C. 1100
- D. 300–500
- E. 1220

REVIEW QUESTIONS 12

Machines and Accessories

True-False

T F **1.** Shielded metal-arc welding is sometimes called arc welding.

T F **2.** Arc heat generated when using the shielded metal-arc process ranges from 6,000°F to 10,000°F.

T F **3.** The actual voltage used to provide the welding current ranges from 220–440 volts.

T F **4.** Reverse polarity is sometimes called direct current positive.

T F **5.** AC has straight polarity only.

T F **6.** Constant current welding machines are used primarily for the shielded metal-arc welding process.

T F **7.** The transformer-type welding machine produces AC current.

T F **8.** Pant cuffs are useful to protect the welder's shoes from molten metal.

T F **9.** A welding machine is rated according to its approximate amperage capacity at 60% duty cycle.

T F **10.** The rectifier-type welding machine is used for DC welding only.

T F **11.** A clear plastic cover plate protects the colored lens in the welding helmet.

Multiple Choice

______ **1.** Which of the following is not true about a constant current welding machine?
- A. It is used primarily for manual stick welding.
- B. It has a sloping volt-amp characteristic.
- C. Current is produced with little change caused by arc length.
- D. The machine is commonly used for gas metal-arc welding.
- E. It permits the shielded metal-arc welder to maintain good control of the puddle.

______ **2.** The transformer-type of welding machine ______.
- A. is the lightest of all welding machines
- B. is the least expensive of all welding machines
- C. produces alternating current
- D. does not produce arc blow
- E. all of the above

______ **3.** Which of the following is not true about the generator-type of welding machine?
- A. It is used primarily for manual stick welding.
- B. It can be used with DC reverse polarity.
- C. It can be used with DC straight polarity.
- D. When it is used with a dual control machine, a soft or harsh arc can be produced.
- E. It can be adjusted for AC current.

KNOW YOUR WELDING SYMBOLS

U-Groove Weld

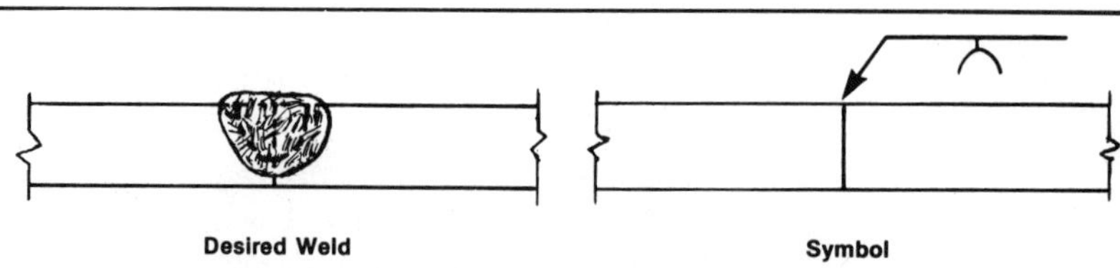

______ 4. ______ is/are required for shielded metal-arc welding.

A. Goggles
B. Gloves
C. An apron
D. A helmet
E. all of the above

______ 5. What is the meaning of the welding symbol shown?

A. V-groove both sides
B. Bevel-groove arrow side
C. V-groove other side
D. U-groove other side
E. none of the above

Matching

______ 1. A(n) ______ allows free passage of electrical current.

______ 2. Electrical current that flows in one direction only is ______.

______ 3. The ______ is the path taken by electrical current.

______ 4. Electrical current with positive and negative values is ______.

______ 5. The ______ is the amount or rate of current that flows in a circuit.

A. alternating current
B. direct current
C. conductor
D. electrical circuit
E. ampere

______ 6. The generation of a stable voltage regardless of the amperage output is called ______.

______ 7. Electricity at rest is ______.

______ 8. A ______ is the force that causes current to move.

______ 9. The opposition of the material in a conductor to the electrical passage of an electrical current is called ______.

______ 10. Electricity in motion in an electrical circuit is called ______.

A. volt
B. resistance
C. static electricity
D. dynamic electricity
E. constant potential

______ 11. Voltage produced when no welding is being done is ______.

______ 12. Electrode positive and work negative results in ______.

______ 13. Electrode negative and work positive results in ______.

______ 14. A ______ is caused by increased distance of work from the welding machine.

______ 15. The ______ control spans a range of voltages in setting the open-circuit voltage on a welding machine.

A. voltage drop
B. open circuit voltage
C. variable voltage
D. straight polarity
E. reverse polarity

______	**16.** A ______ is recommended when welding from 75 amps to 200 amps.	A. eye protection
______	**17.** Suitable ______ must be worn under the welding helmet for protection from slag and other particles.	B. #10 shade
______	**18.** A ______ is used to observe a welder.	C. #12 shade
______	**19.** A ______ is recommended when welding over 400 amps.	D. #14 shade
______	**20.** A ______ is recommended when welding from 200 to 400 amps.	E. hand-held helmet

______	**21.** A ______ is used to remove slag.	A. cables
______	**22.** Welding ______ carry current to and from the work.	B. ventilation
______	**23.** The ______ is connected to the work or bench.	C. chipping hammer
______	**24.** The ______ should be protected by insulation.	D. electrode holder
______	**25.** Removal of welding fumes requires proper ______.	E. ground connection

REVIEW QUESTIONS 13

Selecting Electrodes

True-False

T F **1.** The E-7024 electrode is useful for overhead butt joints.

T F **2.** The diameter of an electrode should be approximately one-half the thickness of the plate to be welded.

T F **3.** Moisture can damage the coating on an electrode.

T F **4.** An electrode should always be used until it is burned down to a 4″ stub.

T F **5.** Some electrodes are designed to be used only with direct current.

T F **6.** Fast-fill electrodes are generally used in situations where all the work can be positioned for flat welding.

T F **7.** Drying ovens are often used to store electrodes, particularly those vulnerable to moisture, such as the E-7018.

T F **8.** DC-EN is the same as DCR (direct current reverse polarity).

T F **9.** Fast-freeze electrodes are best suited for vertical and overhead welding.

T F **10.** The weld metal deposited should have approximately the same mechanical properties that the parent metal has.

T F **11.** Some electrodes have coatings that include iron powder, which converts to steel and becomes part of the weld deposit.

T F **12.** Coatings on some electrodes release oxygen and nitrogen to protect the weld metal.

T F **13.** An E-6010 electrode produces a flat or concave bead using DCR current.

T F **14.** When welding in the vertical and overhead positions, a smaller diameter electrode than that used in the flat position should be used.

T F **15.** Damaged coatings on electrodes do not affect the performance of the electrode.

Multiple Choice

__________ **1.** A function of the coating on a shielded electrode is that it __________.

A. provides easy arc starting
B. protects the weld from atmospheric contamination
C. acts as a cleansing agent
D. permits better penetration
E. all of the above

__________ **2.** __________ is not a consideration in selecting an electrode.

A. Joint design and fit-up
B. Electrode diameter
C. Welding current and polarity
D. Welding position
E. none of the above

KNOW YOUR WELDING SYMBOLS

Flare-V-Groove Weld

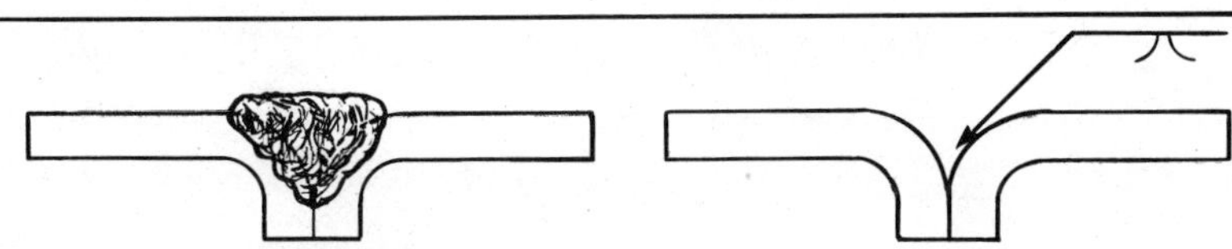

Desired Weld — Symbol

______ 3. What is the meaning of the welding symbol shown?

A. Flare bevel groove other side
B. Flare V-groove arrow side
C. Single V-groove
D. Bevel-groove
E. none of the above

Matching

Match the welding current with the electrode shown.

______ 1. AC, DCR
______ 2. DCR
______ 3. AC, DCS
______ 4. AC, DCR, DCS

A. E-6010
B. E-6011
C. E-6012
D. E-6013

Match the characteristics with the electrode shown.

______ 5. E-7018
______ 6. E-6012
______ 7. E-6010
______ 8. E-7024

A. fast freeze
B. fill-freeze
C. fast-fill
D. low hydrogen

______ 9. The ______ group produces a snappy arc with little slag.
______ 10. The ______ group includes the heavy coated, iron powder electrodes.
______ 11. The ______ group includes general purpose electrodes.

A. fast-freeze
B. fill-freeze
C. fast-fill

Identify the parts of the AWS electrode classification

______ 12. Welding position
______ 13. Special manufacturer's characteristics
______ 14. Tensile strength
______ 15. Electric welding

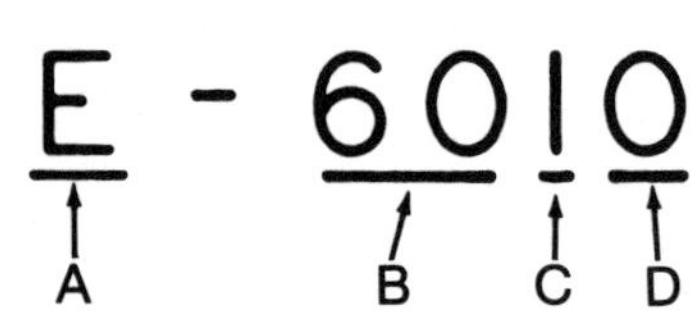

Identify the parts of the shielded metal-arc welding process shown.

______ 16. Solidified metal and electrode
______ 17. Molten metal
______ 18. Arc
______ 19. Wire
______ 20. Slag

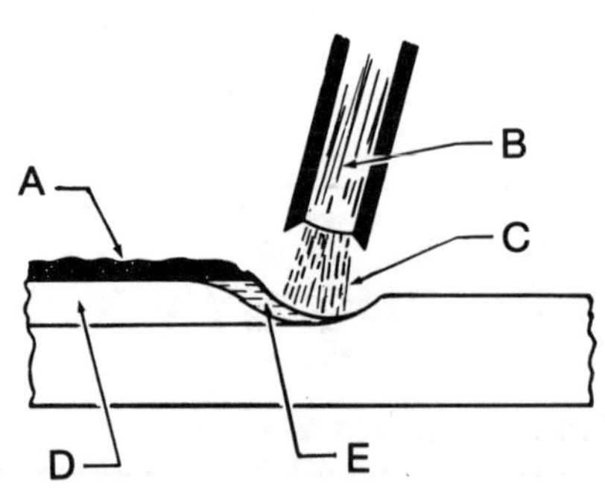

REVIEW QUESTIONS 14

Striking Arcs

True-False

T F 1. The jaws of an electrode holder should always be kept clean to ensure good electrical contact.

T F 2. The recommended amperage range for a specific electrode is approximate and must be adjusted as necessary to achieve a satisfactory weld.

T F 3. Factors such as size and type of electrode, current, position, and amperage all affect the speed of travel.

T F 4. The travel angle can be adjusted by removing the electrode.

T F 5. An electrode welded fast to the plate should be broken loose by twisting or bending the electrode.

T F 6. An arc that is too long will produce a humming sound.

T F 7. By increasing the amperage, an electrode can be made to burn off more slowly.

T F 8. When welding is complete, the machine should be shut off before leaving the work area.

T F 9. Arc length should be maintained at approximately $1/8''$.

T F 10. Welding with too much amperage will cause the electrode to stick.

Matching

_________ 1. The _________ is the angle at which the electrode is held.

_________ 2. The _________ is the distance between the electrode and the work.

_________ 3. The _________ is the rate at which an electrode moves across the weld area.

_________ 4. The _________ is the adjustment of a welding machine for a specific welding operation.

_________ 5. The _________ is the actual flow of electricity regulated by a power supply.

A. machine setting
B. current
C. arc length
D. electrode angle
E. speed of travel

_________ 6. The _________ is gripped in the electrode holder.

_________ 7. Using a(n) _________ increases weld spatter, producing flat wide beads.

_________ 8. The _________ determines the type of electrode to be used.

_________ 9. The _________ will appear clear if no slag is mixed in with the puddle.

_________ 10. Not enough heat is created to melt the base metal when using a(n) _________.

A. molten metal of the weld
B. bare end of the electrode
C. arc length too long
D. arc length too short
E. type of current used

KNOW YOUR WELDING SYMBOLS

Flare-Bevel Weld

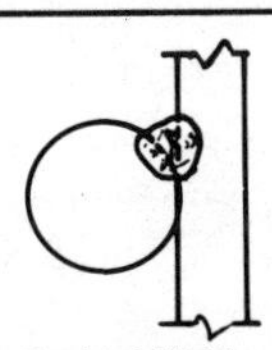

Desired Weld

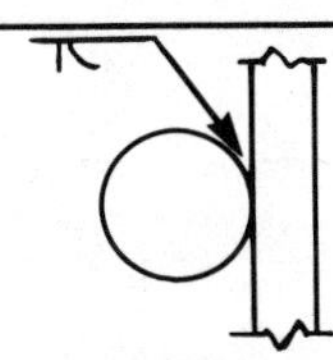

Symbol

__________ **11.** A __________ is used to mark metal for cutting or welding.

__________ **12.** When using the ________, an arc is started by bringing the electrode straight downward.

__________ **13.** A ____________ indicates correct current setting and arc length.

__________ **14.** A ________ is established when the electrode is angled slightly toward the end of the weld when running a continuous bead.

__________ **15.** When using the ________, the arc is started by bringing the electrode in contact with the plate similar in motion to striking a match.

A. scratching motion
B. tapping
C. travel angle
D. crackling noise
E. soapstone

REVIEW QUESTIONS 15

Running Continuous Beads

True-False

T F 1. The length of an arc is determined by the diameter of the electrode.

T F 2. The weld bead will be narrow and the ripples will be pointed if the speed of travel is too slow.

T F 3. Running with an arc length too short increases the possibility of the electrode sticking to the work.

T F 4. Whipping the electrode allows better heat control in the weld puddle when using an E-6010 or E-6011 electrode.

T F 5. Weaving motions frequently used include the crescent, figure 8, and rotary.

T F 6. Padding or surfacing builds up the surface of the base metal.

T F 7. A specific welding technique may be required, depending on the type of electrode used.

T F 8. Welding with too much amperage results in overlapping.

Multiple Choice

_________ 1. Arc length should be as long as _________.
- A. $^{5}/_{16}''$
- B. the uncoated portion of the electrode
- C. one-third to one-half the total thickness of the bead
- D. the diameter of the electrode
- E. none of the above

_________ 2. When cleaning slag from a weld, which of the following is the correct procedure?
- A. Chip the slag away from your body to prevent slag from flying up into your face.
- B. Do not pound the bead too hard.
- C. After the slag is loosened, drag the point end of a hammer along the weld.
- D. Brush the remaining particles of slag with a stiff wire brush.
- E. all of the above

_________ 3. What is the meaning of the welding symbol shown?

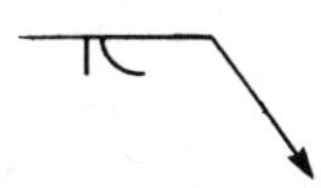

- A. Butt weld arrow side
- B. Bevel-groove weld arrow side
- C. Flare-bevel-groove weld arrow side
- D. Flare-V-groove arrow side
- E. none of the above

KNOW YOUR WELDING SYMBOLS

Plug or Slot Weld

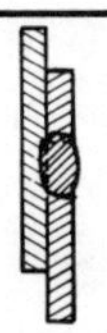

Desired Weld

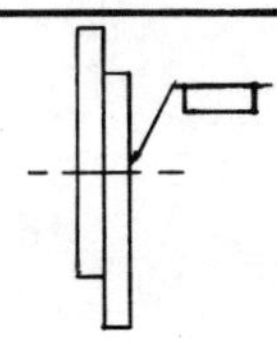

Symbol

________ **4.** What is the meaning of the welding symbol shown?

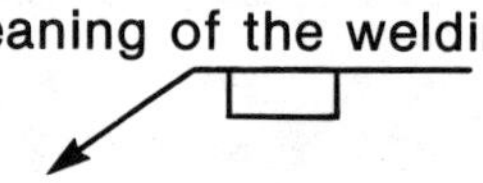

A. Square butt joint
B. Plug or slot weld other side
C. Bevel groove other side
D. Plug or slot weld arrow side
E. none of the above

Matching

________ **1.** The weld bead will show narrow pointed ripples if the incorrect ________ is used.
________ **2.** The properties of the base metal determines the ________ to use.
________ **3.** The ________ may vary from 5° to 30° from the vertical in the line of the welding.
________ **4.** If the ________ is too high the electrode will melt too fast.
________ **5.** The ________ should be approximately 1/8″ when using an electrode 1/8″ in diameter.

A. electrode
B. arc length
C. amperage
D. speed of travel
E. electrode angle

________ **6.** A(n) ________ motion is used to control the temperature of the molten puddle.
________ **7.** When the amperage is too low, ________ occurs.
________ **8.** A(n) ________ forms if the arc comes in contact with the base metal.
________ **9.** The ________ technique is used to increase bead width.
________ **10.** A welding amperage too high results in ________.

A. crater
B. whipping
C. undercutting
D. overlapping
E. weaving

________ **11.** The process of building worn surfaces or shafts is done by ________.
________ **12.** The weld ________ should be approximately one-third to one-half the total thickness of the bead.
________ **13.** The ________ is measured at right angles to the line of welding.
________ **14.** A ________ is necessary after arc is broken to continue the weld.
________ **15.** The amount of ________ used controls the amount of heat directed to the base metal.

A. work angle
B. padding
C. penetration
D. amperage
E. restart

Identify the following:

________ **16.** Crater
________ **17.** Bead thickness
________ **18.** Electrode
________ **19.** Penetration

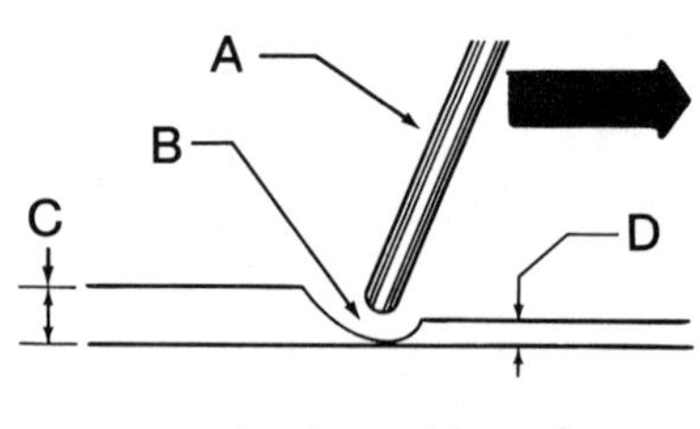

REVIEW QUESTIONS 16

Flat Position

True-False

T F 1. One or more filler passes may be necessary to properly complete a groove weld.

T F 2. A weaving motion is often used when depositing a root pass.

T F 3. When the thickness of the metal exceeds 1/8″, the edges of a butt joint must be beveled.

T F 4. In a multiple pass T-fillet joint, the first pass is welded with a 45° work angle.

T F 5. Slag from the previous pass must be removed completely before welding on the next pass.

T F 6. When welding a lap joint in the flat position, the tacked pieces should be rested against a firebrick.

T F 7. A 90° work angle is used when welding a butt joint in the flat position.

T F 8. A single pass is sufficient when welding a closed butt joint on thin stock.

T F 9. A root pass should not penetrate the bottom surface of the groove more than 1/16″.

T F 10. Welding in the flat position is easier than welding in other positions.

Multiple Choice

__________ 1. A weld that requires more than one layer is called a __________ weld.

A. surface
B. root
C. backstep
D. intermittent
E. multiple pass

__________ 2. A closed butt joint __________.

A. is suitable for welding plates 1/4″ or thicker
B. is used on metal that is 3/16″ or less in thickness
C. is usually welded with three passes
D. allows more penetration of the weld metal than the open butt joint
E. all of the above

__________ 3. The included angle of a V-butt joint should not exceed __________°.

A. 45
B. 90
C. 60
D. 120
E. none of the above

KNOW YOUR WELDING SYMBOLS

Spot Weld

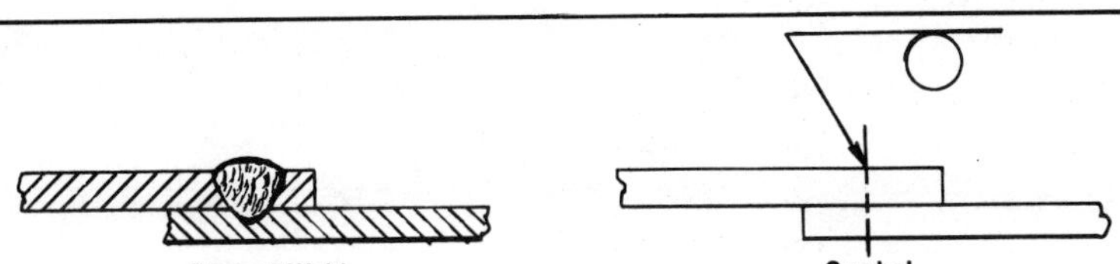

Desired Weld Symbol

______ **4.** When welding a butt joint, a ______° travel angle is used.
A. 90
B. 60
C. 45
D. 70
E. none of the above

______ **5.** What is the meaning of the welding symbol shown?

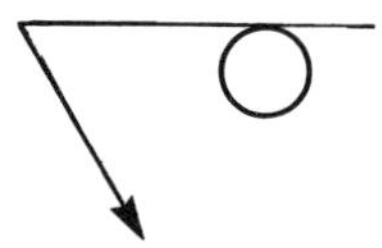

A. Plug weld arrow side
B. Spot weld
C. Butt weld
D. Bevel-groove
E. none of the above

Matching

______ **1.** A ______ produces complete penetration to the bottom surface of the groove.
______ **2.** A ______ is used to provide reinforcement and a good appearance.
______ **3.** A ______ is used to fill the groove of the weld.
______ **4.** A ______ is used to hold the weld pieces in position before other passes are completed.

A. tack weld
B. root pass
C. filler pass
D. cover pass

______ **5.** A(n) ______ is frequently used to fabricate straight and rolled shapes.
______ **6.** No beveling or machining is necessary when using a(n) ______.
______ **7.** A(n) ______ is used to weld round shafts and rods.
______ **8.** A(n) ______ is used to construct rectangular shaped objects, such as tanks.
______ **9.** Opposite sides of the stock should be beveled when welding ______.

A. lap joint
B. T-fillet
C. outside corner
D. butt joint
E. round stock

Identify the parts of the V-butt joint shown.

______ **10.** Root pass
______ **11.** Cover pass
______ **12.** Filler pass
______ **13.** Tack weld

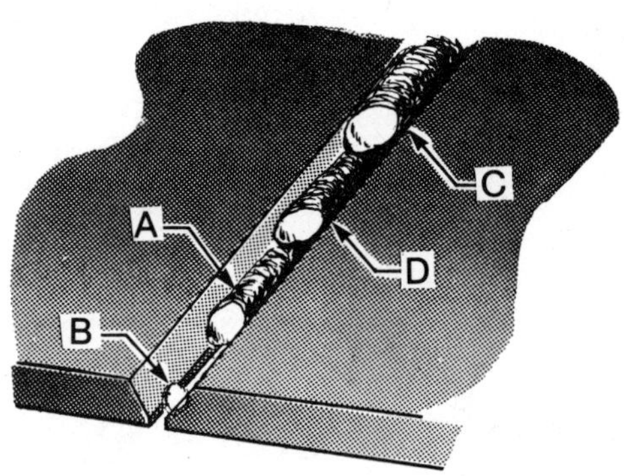

REVIEW QUESTIONS 17

Horizontal Position

True-False

T F 1. Weaving the electrode when welding in the horizontal position allows better heat control.

T F 2. More amperage is required when welding in the horizontal position.

T F 3. A 35° work angle is used when depositing a stringer bead.

T F 4. Maintaining a shorter arc length increases puddle control.

T F 5. The electrode is positioned downward when welding a butt joint in the horizontal position.

Multiple Choice

______ 1. ______ is the recommended current to use when welding with an E-6010 electrode.
- A. AC
- B. DC+
- C. DC−
- D. DC+, DC−
- E. Any polarity

______ 2. When welding in the horizontal position, undercut is caused by ______.
- A. too much amperage
- B. arc length too long
- C. improper electrode work angle
- D. improper electrode motion
- E. all of the above

______ 3. The recommended shade number to use when arc welding at 150 amps is ______.
- A. 4
- B. 5
- C. 6
- D. 8
- E. 10

______ 4. Which of the following electrodes is a fast-freeze type commonly used for welding in the horizontal position?
- A. E-6010
- B. E-6012
- C. E-6013
- D. E-7014
- E. E-7024

KNOW YOUR WELDING SYMBOLS

Seam Weld

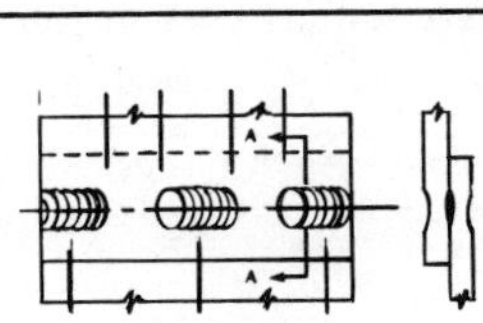

Desired Weld

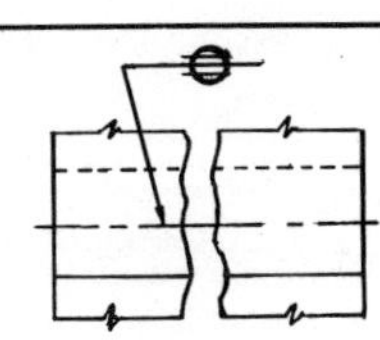

Symbol

_______ 5. A single V-butt welded in the horizontal position commonly has an included angle of _______°.
   A. 37
   B. 45
   C. 60
   D. 75
   E. 120

_______ 6. What is the meaning of the welding symbol shown?

   A. Plug weld
   B. Spot weld
   C. Slot weld
   D. Seam weld
   E. none of the above

Matching

_______ 1. A _______° work angle of the electrode is used when depositing the first pass of a multiple-pass T-fillet.

_______ 2. A _______° work angle of the electrode is used when depositing the second pass of a multiple-pass T-fillet.

_______ 3. A _______° work angle of the electrode is used when depositing the third pass of a multiple-pass T-fillet

_______ 4. A _______° travel angle of the electrode is used when depositing stringer beads in the horizontal position.

_______ 5. A _______° work angle of the electrode is used when depositing stringer beads in the horizontal position.

A. 45
B. 30
C. 10
D. 70
E. 20

_______ 6. A(n) _______ is a welding defect caused by molten metal running down from the bead and solidifying without penetrating.

_______ 7. The deepest part of the weld joint is the _______.

_______ 8. A(n) _______ occurs when a crater is not filled by metal deposited by the electrode.

_______ 9. The _______ type of electrode is best suited for horizontal welding.

_______ 10. A(n) _______ is used to produce a smooth finish on the weld metal.

A. fast-freeze
B. overlap
C. cover pass
D. root
E. undercut

REVIEW QUESTIONS 18

Vertical Position

True-False

T F 1. Fast-freeze and fill-freeze electrodes are recommended for welding in the vertical position.

T F 2. The arc length should be kept short when vertical welding.

T F 3. Whipping the electrode when welding vertical up allows for more control of the weld puddle.

T F 4. When practicing welding in the vertical position, amperage should be increased to levels higher than those used in the flat position.

T F 5. A wash bead should be used on the final pass if a smooth weld is required.

T F 6. A whipping motion is used with an E-7018 electrode.

T F 7. When welding vertical up, the arc should be struck at the top of the plate.

T F 8. The width of the bead can be increased by using a weaving motion.

T F 9. When welding a butt joint vertical up, the cover pass should penetrate deep into the root of the weld.

T F 10. The slag between each pass should be carefully removed to ensure a sound weld.

T F 11. A light dragging motion with a short arc is used when welding vertical down with an E-7018 electrode.

T F 12. When using a whipping motion, the arc should be broken so the metal may cool.

Multiple Choice

________ 1. A ________° travel angle is used when welding vertical down.
- A. 10–15
- B. 5
- C. 10
- D. 15–30
- E. none of the above

________ 2. Which of the following is not true about SMAW vertical down welding?
- A. Less penetration is achieved compared to vertical up.
- B. It is recommended for lighter materials.
- C. It can produce a weld faster than vertical up.
- D. It is used for metals ¼″ and greater in thickness.
- E. all of the above

________ 3. Which of the following is not true about SMAW vertical up welding?
- A. It is recommended for thicker metals.
- B. A whipping motion is recommended when using fast-freeze electrodes.
- C. A whipping motion is recommended to increase the puddle width.
- D. A 10°–15° travel angle is used.
- E. none of the above

KNOW YOUR WELDING SYMBOLS

Back Weld

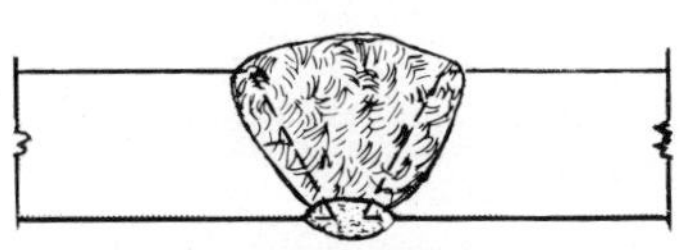

Desired Weld

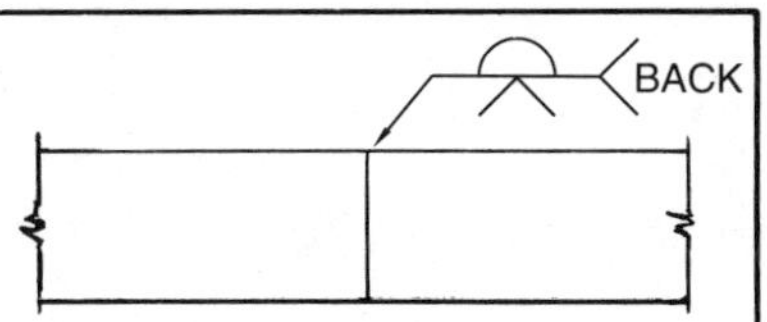

Symbol

_______ **4.** When running straight beads vertical up, _______ when whipping the electrode.
  A. use a fast-freeze electrode
  B. deposit beads with a drag motion
  C. break the arc to allow the puddle to cool
  D. hold the arc in a stationary position
  E. none of the above

_______ **5.** A _______ is used when a smooth weld is required on the final pass of a wide joint.
  A. whipping motion
  B. wash pass
  C. root pass
  D. filler pass
  E. none of the above

_______ **6.** What is the meaning of the welding symbol shown?

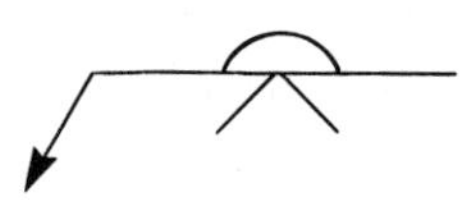

  A. Groove weld with backing
  B. Fillet weld
  C. Groove weld with slot
  D. Bevel-groove with backing
  E. none of the above

Matching

Identify the following passes:

_______ **1.** Filler
_______ **2.** Cover
_______ **3.** Root

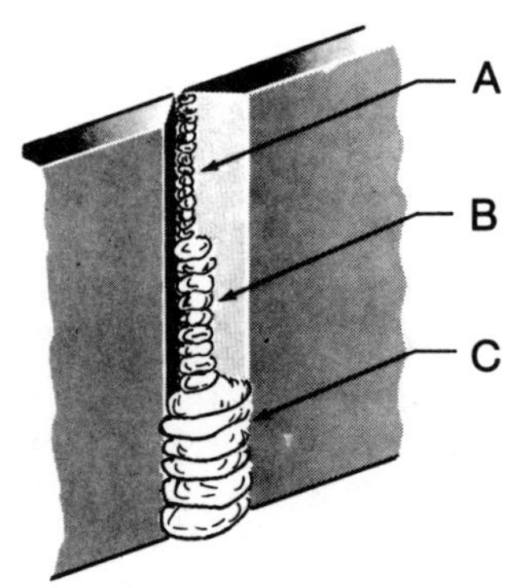

REVIEW QUESTIONS 19

Overhead Position

True-False

T F **1.** Of the four welding positions, welding in the overhead position is the most difficult to master.

T F **2.** When welding in the overhead position, amperage should be reduced as necessary.

T F **3.** A lap joint welded in the overhead position requires a work angle of 45° on the first pass.

T F **4.** Fast-freeze electrodes are recommended with the proper motion when welding in the overhead position.

T F **5.** A positioner is used to secure the weld pieces in the correct position for welding in the overhead position.

T F **6.** The edges of a single V-butt joint should be beveled when welding in an overhead position.

T F **7.** A 75° work angle is used when running beads in the overhead position.

Multiple Choice

________ **1.** A ________° travel angle should be used when welding in the overhead position.

A. 10–15
B. 30
C. 45
D. 60
E. none of the above

________ **2.** Which of the following is not true about welding in the overhead position?

A. Grip the electrode holder so the knuckles are up and the palm is down.
B. Drape the welding cable over your shoulder if welding in a standing position.
C. Stand to one side of the arc rather than underneath it.
D. Increase the arc length to obtain maximum penetration.
E. Use both hands to hold the electrode holder if necessary.

________ **3.** What safety precaution should be followed when welding in the overhead position?

A. Wear a welding cap
B. Wear a garment with a tight-fitting collar.
C. Roll down sleeves and pant cuffs.
D. all of the above
E. none of the above

KNOW YOUR WELDING SYMBOLS

Melt-Thru-Weld

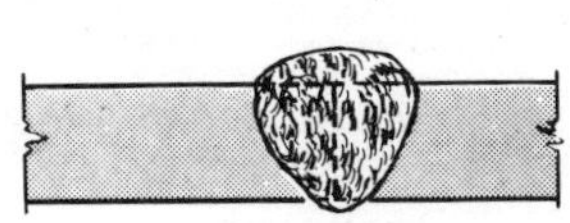

Desired Weld

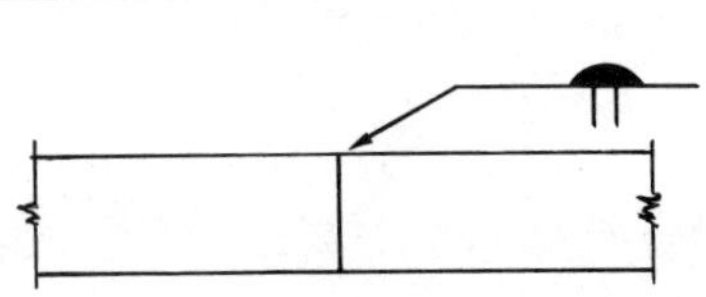

Symbol

__________ **4.** __________ should be kept as short as possible when welding running beads in the overhead position.

A. Cable length
B. Electrode
C. Weld bead
D. Arc length
E. none of the above

__________ **5.** A work angle of __________° is recommended when welding a lap joint in the overhead position.

A. 15
B. 90
C. 75
D. 45
E. none of the above

__________ **6.** What is the meaning of the welding symbol shown?

A. Square butt with melt thru
B. Bevel-groove with no joint preparation
C. Flare-groove with melt-thru
D. Double-groove weld
E. none of the above

REVIEW QUESTIONS 20

Cast Iron

True-False

T F 1. Cast iron should never be heated beyond a dull red color or a temperature exceeding 1,200°F.

T F 2. Studs are used to strengthen a joint when repairing castings greater than 1½″ thick.

T F 3. Generally, DCS (direct current straight) polarity is used for welding cast iron.

T F 4. After the weld is deposited, the entire casting should be peened to relieve any stresses that have been built up.

T F 5. An included angle of approximately 60° is used when preparing a single V-butt joint.

T F 6. Cast iron has an iron base and contains a low percentage of carbon.

T F 7. Spark characteristics on a grinder can be used to identify the type of cast iron.

Multiple Choice

__________ 1. Which of the following procedures is not followed when welding cast iron?
- A. Keep the casting as cool as possible.
- B. Carefully clean the casting before welding.
- C. Maintain an amperage setting higher than that for mild steel.
- D. Preheat the piece whenever possible.
- E. all of the above

__________ 2. Which of the following is true about preheating the casting before welding.
- A. A temperature stick can be used to determine correct temperature.
- B. The preheat temperature should be uniform over the entire casting.
- C. The preheat temperature should be approximately 500°-1200°F.
- D. The preheat temperature should be maintained until the weld is completed.
- E. all of the above

__________ 3. Which of the following is not true about welding cracks in cast iron?
- A. V the crack approximately 1/8″-3/16″ deep.
- B. Cracks in the casting can be made more visible by rubbing a piece of chalk over the surface.
- C. The joint should be ground to remove the casting skin.
- D. A backstep welding technique should be used when necessary.
- E. Large diameter electrodes should be used for quick deposits.

__________ 4. A ____________ welding technique is used to prevent cracks from forming in the casting.
- A. whipping motion
- B. back-step
- C. 2″ weaving motion
- D. all of the above
- E. none of the above

KNOW YOUR WELDING SYMBOLS

Surfacing Weld

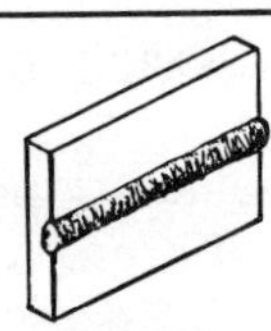

Desired Weld

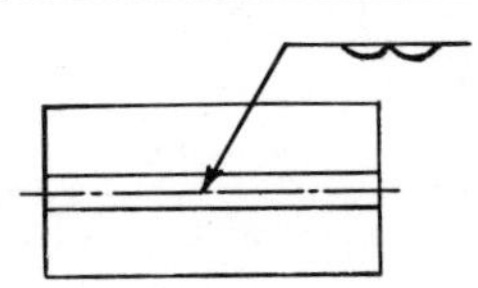

Symbol

______ **5.** ______ electrodes are used to keep heat to a minimum.
- A. DC straight
- B. Large diameter
- C. Small diameter
- D. One-eighth inch diameter or greater
- E. all of the above

______ **6.** ______ are used to reinforce the weld when repairing a casting $1\frac{1}{2}$″ or more in thickness.
- A. Screws
- B. Butt joints
- C. Studs
- D. Square butt joints
- E. none of the above

______ **7.** Most electrodes designed for welding cast iron use ______ current.
- A. AC and DCR
- B. DCR and DCS
- C. AC and DCS
- D. all of the above
- E. none of the above

______ **8.** What is the meaning of the welding symbol shown?

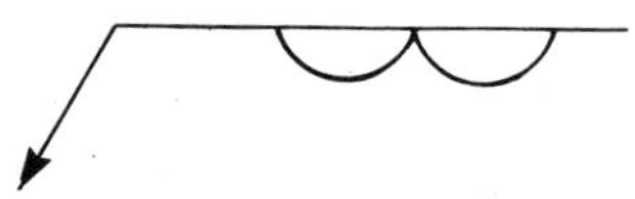

- A. Back weld
- B. Spot weld
- C. Surfacing weld
- D. Fillet weld
- E. none of the above

Matching

______ **1.** ______ is white cast iron that has been subjected to a long annealing process.

______ **2.** ______ is sometimes called ductile iron.

______ **3.** ______ contains combined carbon.

______ **4.** ______ contains elements such as copper, aluminum, nickel, and titanium.

______ **5.** ______ results when the silicon content is high and the iron is permitted to cool slowly.

- A. Gray cast iron
- B. White cast iron
- C. Malleable cast iron
- D. Alloy cast iron
- E. Nodular cast iron

______ **6.** ______ relieves stresses in the weld area.

______ **7.** ______ deposits are soft and ductile.

______ **8.** ______ deposits are hard and waterproof.

______ **9.** ______ prevents a crack in a casting from spreading.

______ **10.** ______ allows necessary penetration of the weld into the weld area.

- A. Machinable electrode
- B. Nonmachinable electrode
- C. Peening the weld
- D. V-ing the crack
- E. Drilling a $\frac{1}{8}$″ hole

REVIEW QUESTIONS 21

Carbon Steels

True-False

T F 1. A postweld heat treatment is recommended for many carbon steels.

T F 2. The E-60XX series of electrodes can be used to weld low-carbon steels.

T F 3. Porosity in a weld is caused from excess carbon deposited by the electrode.

T F 4. Cracks in the weld metal often are invisible to the naked eye.

T F 5. The strength of high-carbon steel depends upon its hardness, which is obtained by a heat-treating process.

T F 6. The control heat process builds up heat in the base metal by using high welding amperage and slow travel speed to eliminate the need for preheating.

Matching

______ 1. ______ steels are considered more difficult to weld than other carbon steels.

______ 2. ______ steels receive no deoxidizing treatment.

______ 3. ______ steels are generally the easiest steels to weld.

______ 4. ______ steels have a smooth surface and contain no blowholes.

______ 5. ______ steels contain 0.30%–0.45% carbon.

A. Low-carbon
B. Medium-carbon
C. High-carbon
D. Killed
E. Rimmed

______ 6. ______ appear when fillet welds are improperly formed.

______ 7. ______ is caused by gas trapped in the weld metal.

______ 8. ______ involves heating the base metal before welding.

______ 9. ______ includes temperatures for stress relief ranging from 900°F–1,250°F.

______ 10. ______ are caused by excessive carbon in the first layer of weld metal.

A. Preheating
B. Postweld heat treatment
C. Crater cracks
D. Root cracks
E. Porosity

______ 11. Welding heat has no appreciable effect on the parent metal when welding ______.

______ 12. Generally, preheating and postweld heat treatment are advised when welding ______.

______ 13. Welding ______ is relatively easy when using E-70XX electrodes.

A. low-carbon steel
B. medium-carbon steel
C. high-carbon steel

KNOW YOUR WELDING SYMBOLS

Edge Flange

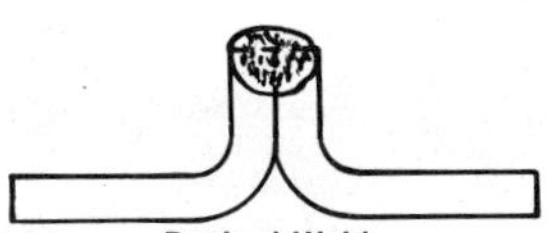
Desired Weld

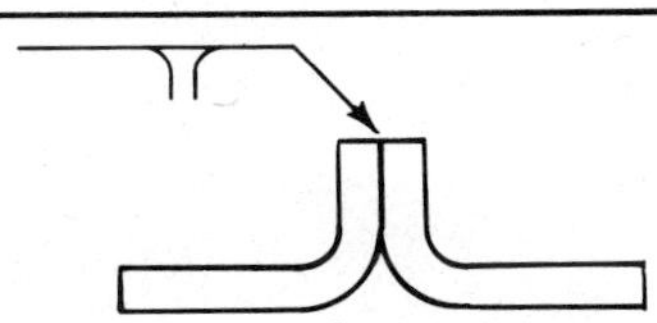
Symbol

Multiple Choice

_______ 1. Which of the following methods can be used to determine the correct preheat temperature?
   A. Rub 50-50 solder on the surface.
   B. Use a surface thermometer.
   C. Mark the surface with carpenter's chalk.
   D. Use a temperature indicating crayon.
   E. all of the above

_______ 2. Which of the following is true when correctly preheating the base metal?
   A. It lessens the chances of martensite formation.
   B. It burns grease, oil, and scale out of the joint.
   C. Its main purpose is to slow the cooling rate of the weld.
   D. It reduces distortion.
   E. all of the above

_______ 3. What is the meaning of the welding symbol shown?

   A. Flare-groove
   B. Flare-bevel
   C. Edge-flange
   D. Single-J
   E. none of the above

REVIEW QUESTIONS 22

Alloy Steels

True-False

T F **1.** An alloy steel is a steel mixed with one or more additional elements.

T F **2.** Preheating and postweld heat treatment are necessary for some alloy steels.

T F **3.** Stainless steel is classified in the AISI 200–300 and 400 series.

T F **4.** When welding stainless steel, chill plates are not recommended for reducing unfavorable effects of heat.

T F **5.** Inert-gas shielded welding is preferred when welding stainless steel because of its reduced heat input.

T F **6.** Welding clad steel requires two different electrodes.

T F **7.** When welding stainless steel, a long arc length should be used to obtain maximum penetration.

T F **8.** Alloy steels often require electrodes different from those used for mild steel.

T F **9.** The first stainless steel developed was the chromium type.

T F **10.** At least 20% more current is recommended for welding stainless steel than for welding mild steel.

Matching

_______ **1.** A(n) _______ is used for parts that must withstand impact stresses.

_______ **2.** The properties of _______ include strength, toughness, and corrosion resistance.

_______ **3.** Other names for _______ are carbon-moly and chrome-moly.

_______ **4.** A(n) _______ determines the correct preheat procedures.

_______ **5.** A tough, nonmagnetic alloy noted for its high strength is _______.

A. clip test
B. austenitic manganese steel
C. low-manganese steel
D. low-alloy molybdenum steel
E. low-alloy nickel steel

Identify the chemical composition of electrodes containing the following letters in the electrode classification suffix.

_______ **6.** Suffix A
_______ **7.** Suffix B
_______ **8.** Suffix C
_______ **9.** Suffix D
_______ **10.** Suffix G

A. manganese-molybdenum
B. carbon-molybdenum
C. chromium-molybdenum
D. nickel steel
E. other low-alloy electrodes with minimum elements

KNOW YOUR WELDING SYMBOLS

Corner Flange

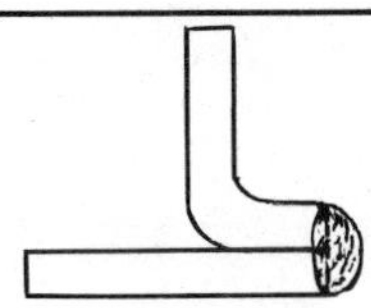

Desired Weld

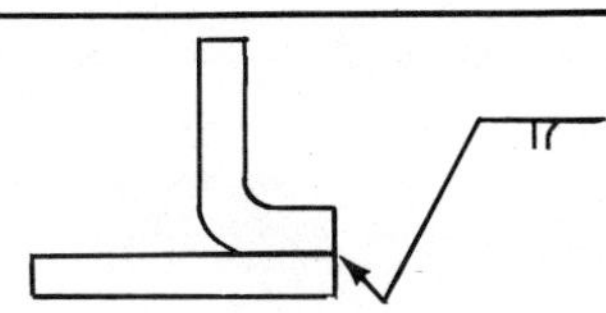

Symbol

Identify the procedure for welding clad steel shown.

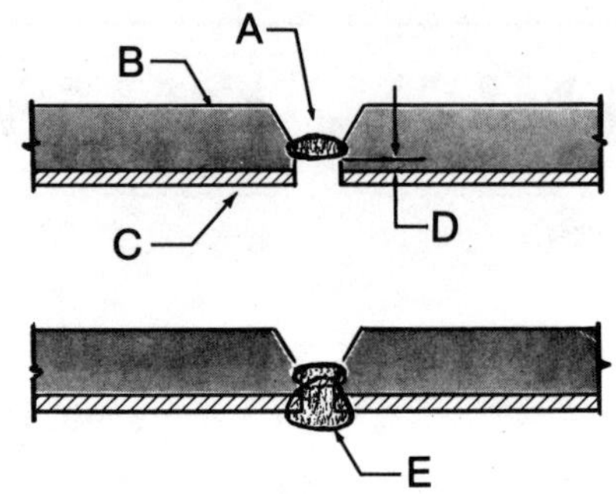

_________ **11.** Depth of penetration

_________ **12.** Root pass - mild steel electrode

_________ **13.** Cladding

_________ **14.** Mild steel

_________ **15.** Alloying electrode

Multiple Choice

_________ **1.** What is the meaning of the welding symbol shown?

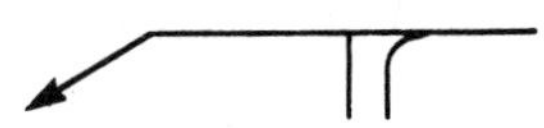

A. Flare-bevel
B. Corner-flange
C. Butt
D. Flare-groove
E. none of the above

_________ **2.** What is the meaning of the welding symbol shown?

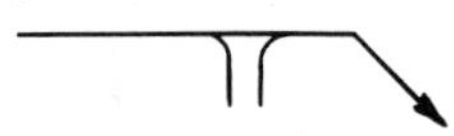

A. Flare-groove
B. Flare-bevel
C. Edge-flange
D. Single-J
E. none of the above

REVIEW QUESTIONS 23

Nonferrous Metals

True-False

T F **1.** Aluminum is less than one-third the weight of other commonly used metals.

T F **2.** Aluminum does not change color when welded.

T F **3.** The oxyacetylene welding process is the method most commonly used for joining nonferrous metals.

T F **4.** Aluminum in its pure form does not possess a very high tensile strength.

T F **5.** Steel is a common alloying ingredient in nonferrous metals.

T F **6.** Nonheat-treatable alloys are usually designated in the 1000, 3000, 4000, or 5000 series.

T F **7.** Clad alloys contain copper or zinc as the major alloying ingredient.

T F **8.** The edges of aluminum ¼″ thick or thicker should be beveled before welding.

T F **9.** Adequate ventilation or a respirator, if necessary, is required when welding brass.

T F **10.** Monel contains 70% aluminum along with other alloying ingredients.

T F **11.** Deoxidized copper is easier to weld than oxygen-bearing copper.

T F **12.** A scratching motion is recommended when striking an arc on aluminum.

T F **13.** Nonferrous metals require electrodes different from those used on mild steel.

T F **14.** Nonferrous metals include aluminum, copper, brass, bronze, monel, and inconel.

Matching

__________ **1.** __________ contain one or more alloying elements, resulting in a higher tensile strength.

__________ **2.** _________ consists of letters that identify treatments used to produce various tempers.

__________ **3.** __________ do not contain iron.

__________ **4.** _________ are used to produce aluminum that is poured into sand or a permanent mold.

__________ **5.** _______ does not contain any alloying ingredients.

A. Casting alloys
B. Wrought alloys
C. Commercially pure aluminum
D. Temper designation
E. Nonferrous metals

KNOW YOUR WELDING SYMBOLS

Fillet Weld Leg Size

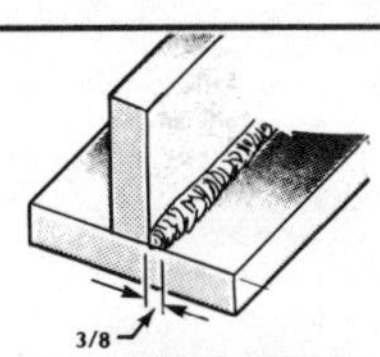

Desired Weld

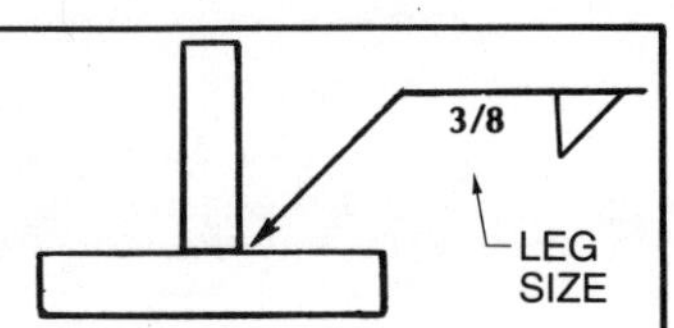

Symbol

_____ 6. _____ is a copper-tin alloy.

_____ 7. _____ is a good conductor of heat and electricity.

_____ 8. _____ is a copper-zinc alloy.

_____ 9. _____ contain more than 50% nickel.

_____ 10. _____ is a soft, ductile metal that hardens when it is cold-worked.

A. Copper
B. Oxyen-bearing copper
C. Brass
D. Bronze
E. Monel and inconel

Identify the following basic temper designations:

_____ 11. F
_____ 12. O
_____ 13. H
_____ 14. W
_____ 15. T

A. as fabricated
B. strength improved by strain hardening
C. thermally treated
D. solution heat-treated
E. fully annealed

Multiple Choice

_____ 1. What is the meaning of the dimensions on the welding symbol shown?

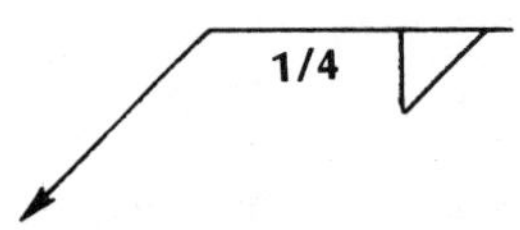

A. Unequal fillet legs
B. Flare-groove ¼″ in length
C. ¼″ Fillet leg size
D. ¼″ Fillet weld length
E. none of the above

REVIEW QUESTIONS 24

Gas Tungsten-Arc Welding—GTAW

True-False

T F **1.** A filler rod is added into the weld with an in-and-out motion.

T F **2.** The electrode is withdrawn quickly from the base metal after the arc is struck.

T F **3.** ACHF stands for alternating current high frequency.

T F **4.** Filler metal is necessary on all welds made with the GTAW process.

T F **5.** DCRP allows deeper penetration than DCSP.

T F **6.** A foot control allows the operator to adjust the amount of current flowing to the work.

T F **7.** Post purge is the amount of time the shielding gas continues to flow after the welding current is stopped.

T F **8.** ACHF welding on aluminum provides deep penetration combined with good cleansing action.

T F **9.** Aluminum has a lower melting point than mild steel.

T F **10.** A notch is formed in the middle of a lap weld as the bead is deposited across the piece.

T F **11.** The GTAW process uses an electrode that is not consumed in the weld.

T F **12.** ACHF is recommended for welding aluminum.

T F **13.** When welding with DCSP, the electrode is positive and the work is negative.

T F **14.** Tungsten electrodes alloyed with thorium last longer and have a higher current capacity than pure tungsten electrodes.

T F **15.** Generally, the diameter of the filler rod should equal the thickness of the metal to be welded.

T F **16.** The addition of helium to argon when used as a shielding gas improves the penetration achieved.

T F **17.** When welding a T-joint, filler metal is required regardless of the thickness of the metal.

T F **18.** On light gauge materials, backing is used to protect the weld from burning through the metal.

T F **19.** The water flow on a water-cooled torch should be turned on after the weld is made to provide adequate cooling.

T F **20.** When using AC current, the electrode does not touch the work when striking the arc.

T F **21.** When welding in the overhead position, the current should be reduced 5–10% of the current used for flat position welding.

T F **22.** In the gas tungsten-arc hot-wire welding process, filler metal is automatically fed from a wire feeder.

T F **23.** Pulsed current GTAW uses two levels of welding current.

KNOW YOUR WELDING SYMBOLS

Single V-Groove Penetration

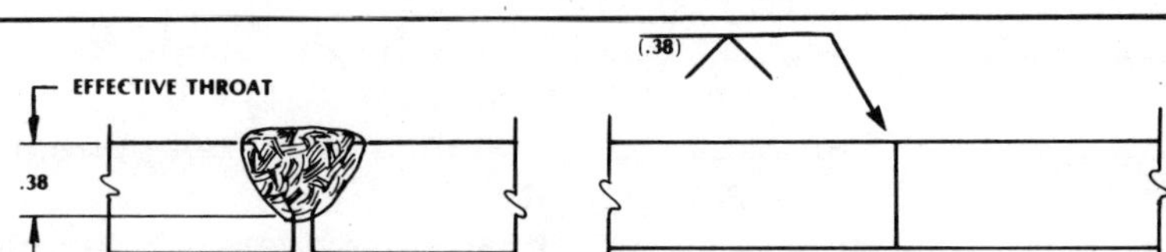

Desired Weld Symbol

T F 24. A push travel angle is most commonly used when using the GTAW process.

T F 25. Toxic fumes are given off when welding copper and copper alloys.

T F 26. High-carbon steels are weldable, but require preheat and postweld heat treatment.

T F 27. To fill the crater at the end of a weld, reduce the amperage using the foot control and continue to add filler metal.

T F 28. Clean the aluminum to be welded with a stainless steel brush before welding.

T F 29. To obtain a spherical end on the electrode, the arc is struck on a copper plate.

T F 30. The electrode should be bent slightly for welding corner joints and lap joints.

T F 31. A small rotary motion is used with the torch when running beads with a filler rod.

Multiple Choice

_________ 1. Which of the following is not true about GTAW?
- A. Direct current straight polarity is often used.
- B. Argon is commonly used as a shielding gas.
- C. The weld is easier to view since the shielding gas is transparent.
- D. Direct current reverse polarity is often used.
- E. none of the above

_________ 2. Which of the following is true when running beads on mild steel using the GTAW process?
- A. 1/8″ to 3/16″ stickout is used.
- B. Amperage is set for 50–60 amps.
- C. A 20° push angle is used.
- D. DCSP is used.
- E. all of the above

_________ 3. What is the meaning of the dimension on the welding symbol shown?

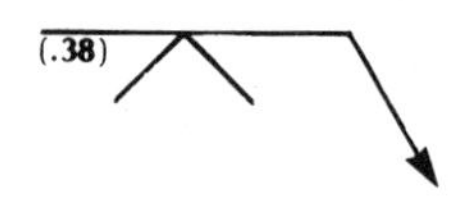

- A. Bevel angle
- B. Leg size
- C. Effective throat size
- D. Back weld size
- E. none of the above

Matching

_________ 1. _________ stands for direct current straight polarity.

_________ 2. _________ stands for alternating current.

_________ 3. _________ is another name for gas tungsten-arc welding (GTAW).

_________ 4. _________ stands for direct current reverse polarity.

_________ 5. _________ is another name for gas metal-arc welding (GMAW).

A. TIG (tungsten inert gas)
B. MIG (metal inert gas)
C. AC
D. DCRP
E. DCSP

_________ **6.** Applying the weld using the _________ process allows the operator to control direction and the speed of travel.

_________ **7.** Weld size, weld length, rate of travel, and starting and stopping of the weld operation are controlled by equipment under the observation of the operator in the _________ process.

_________ **8.** Welding is done by hand in the _________ process.

_________ **9.** Constant observation and adjustment of controls by an operator are not required in the _________ process.

A. manual
B. semiautomatic
C. machine
D. automatic

_________ **10.** _________ is rarely used in GTAW.

_________ **11.** _________ is a combination of DCRP and DCSP.

_________ **12.** When using _________, high frequency is incorporated in the current.

_________ **13.** _________ provides deep penetration.

A. AC
B. DCRP
C. DCSP
D. ACHF

_________ **14.** Argon is most commonly used as a _________.

_________ **15.** A _________ is used to prevent turbulence of the gas stream.

_________ **16.** A _________ is a threaded ceramic nozzle which varies in size.

_________ **17.** The _________ is shaped to a point.

_________ **18.** The _________ is shaped to a spherical end.

A. gas cup
B. gas lens
C. shielding gas
D. AC welding electrode
E. DC welding electrode

_________ **19.** The _________ must have the same composition as the base metal.

_________ **20.** The _________ controls the amount of gas flow in cubic feet per hour.

_________ **21.** The _________ is governed by the current to be used and the thickness of the base metal.

_________ **22.** The _________ allows current to be adjusted easily throughout the weld.

_________ **23.** The _________ control includes a timer that maintains gas flow after weld current is stopped.

A. foot control
B. post purge
C. flowmeter
D. filler rod
E. electrode diameter

_________ **24.** The _________ is designed for welding on light gauge materials at amperages less than 200 amps.

_________ **25.** The _________ is adjusted as necessary for different joints.

_________ **26.** The _________ is designed for welding at amperages of 200 amps and above.

_________ **27.** When using _________, the electrode touches the work when striking the arc.

_________ **28.** A _________ is used when the torch is pointing toward the end of the weld.

A. electrode extension (stickout)
B. water-cooled torch
C. air-cooled torch
D. push angle
E. DC current

REVIEW QUESTIONS 25

Gas Metal-Arc Welding—GMAW

True-False

T F 1. CO_2 has an affinity for argon at higher temperatures.

T F 2. Globular transfer is often used for welding metals thicker than ½″.

T F 3. Nitrogen is one of the elements that make up air.

T F 4. A constant potential welding machine is used for gas metal-arc welding.

T F 5. Unlike GTAW which uses a nonconsumable electrode, GMAW uses a consumable electrode.

T F 6. The welding gun should not be removed from the weld until the weld puddle is solidified.

T F 7. To ascertain the correct welding amperage, start by consulting recommended amperages for the specific welding operation.

T F 8. Voltage increases by increasing the wire feed speed.

T F 9. The GMAW process can be applied by the semiautomatic process only.

T F 10. DCSP is not recommended for gas metal-arc welding.

T F 11. Short-circuit transfer is used most frequently at current levels less than 200 amps.

T F 12. By varying the current and shielding gas, different types of metal transfer occur.

T F 13. DCRP in the GMAW process contributes to better melting, deeper penetration, and excellent cleansing action.

T F 14. When using a constant-current welding machine, voltage changes with arc length.

T F 15. Slope control allows an operator to alter the pinch force of the welding machine.

T F 16. The constant-potential welding machine has a nearly flat volt-ampere curve.

T F 17. The shielding gas in the arc area displaces air with inert gas.

T F 18. Oxygen comprises 78% of the air we breathe.

T F 19. Having the proper amount of shielding gas usually results in a rapidly crackling or sizzling arc sound.

T F 20. E70S-3 filler wire is recommended for low to medium-carbon steels.

T F 21. The edges of metals thicker than ¼″ must be beveled when gas metal-arc welding.

T F 22. Good electrical contact is necessary to start the arc when gas metal-arc welding.

T F 23. Cold laps occur from welding at amperage levels higher than recommended.

T F 24. Thin steel plates ¼″ to ½″ may be welded with square edges.

T F 25. Whiskers in the weld joint are caused by pushing the wire past the leading edge of the puddle.

KNOW YOUR WELDING SYMBOLS

Combined Welding Symbols

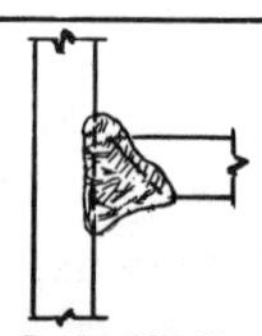

Desired Weld

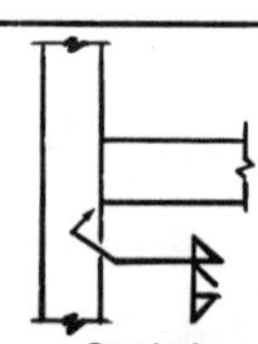

Symbol

T F **26.** To obtain a specific amperage, another person should observe the ammeter while the weld is made.

T F **27.** If the electrode extension (stickout) is increased, welding amperage is decreased.

T F **28.** Alternating current is used for welding in the overhead position on mild steel.

T F **29.** Gas metal-arc welding requires more skill than shielded metal-arc welding.

T F **30.** Usually, the recommended rate of gas flow is 40–60 cfm.

T F **31.** Surface porosity in the weld is caused by insufficient amperage.

T F **32.** A 45° work angle is used when welding the first pass of a multiple-pass T-joint.

Multiple Choice

________ **1.** Which of the following is not true about GMAW?
- A. GMAW is faster than SMAW.
- B. Less time is required to train the operator.
- C. No slag is in the weld deposits.
- D. GMAW has deep penetrating characteristics.
- E. GMAW requires more skill than SMAW.

________ **2.** ________ is most frequently used in the GMAW process.
- A. DCRP
- B. DCSP
- C. AC
- D. ACHF
- E. none of the above

________ **3.** Which of the following elements in the air causes the most severe problems in welding steel?
- A. oxygen
- B. nitrogen
- C. hydrogen
- D. carbon dioxide
- E. none of the above

________ **4.** ________ is the shielding gas most commonly used for steel.
- A. Argon
- B. Helium
- C. Carbon dioxide
- D. Argon and oxygen
- E. none of the above

________ **5.** The shielding gas ________ is not an inert gas.
- A. argon
- B. helium
- C. neon
- D. carbon dioxide
- E. none of the above

________ **6.** ________ is adjusted when using the wire feeder.
- A. Voltage
- B. Shielding gas flow
- C. Polarity
- D. Amperage
- E. none of the above

________ **7.** ________ is another term for stickout.
- A. Nozzle
- B. Gas drift
- C. Electrode extension
- D. Contactor
- E. none of the above

_______ **8.** Which of the following is not true about CO_2?
A. Compared to helium, CO_2 is less expensive.
B. CO_2 is frequently used when welding aluminum.
C. It is not considered an inert gas.
D. Deoxidizers must be added to wires using CO_2 as a shielding gas.
E. all of the above

_______ **9.** GMAW is faster than SMAW because _______.
A. a different polarity is used
B. the shielding gas permits contamination into the weld
C. the slag is easier to remove in the GMAW process
D. a different welding amperage is required
E. electrode change is unnecessary

_______ **10.** The _______ type of GMAW metal transfer is used for welding heavy gauge metals.
A. globular
B. short circuit
C. spray
D. high frequency
E. all of the above

_______ **11.** DCSP is impractical with GMAW because _______.
A. the weld penetration is wide and shallow
B. it produces a good cleansing atmosphere
C. the transfer is in a fine spray
D. weld spatter is minimal
E. none of the above

_______ **12.** Which of the following is true about GMAW spray transfer?
A. A mixture of argon and oxygen is used as a shielding gas.
B. It requires a high current density.
C. Deep weld penetration is possible.
D. Only small droplets are permitted to form.
E. all of the above

_______ **13.** The globular type transfer is used _______.
A. where low heat input welds are desired
B. when welding on metals over ½″ thick
C. where metal is deposited in a spray of fine metal particles
D. for maximum penetration in all positions
E. all of the above

_______ **14.** Which of the following applies to the constant current welding machine?
A. Voltage varies with the length of the arc.
B. It is recommended for GMAW.
C. A preset voltage level is maintained regardless of amperage.
D. The power supply is self-correcting with respect to arc length.
E. all of the above

_______ **15.** Which of the following applies to the constant potential welding machine?
A. When the wire is fed faster, current increases.
B. There is no current control.
C. The machine provides the necessary current required by the load imposed on it.
D. The power supply is self-correcting with respect to arc length.
E. all of the above

_______ **16.** Which of the following is not true about the wire feeder?
A. It can be mounted on the power supply or in a separate location.
B. The wire is driven from the wire spool to the gun and arc.
C. By adjusting a control, the wire speed can be increased or decreased.
D. A solenoid energizes the gas flow to the gun.
E. It has separate controls for voltage, amperage, and wire feed speed.

__________ **17.** Which of the following applies to the GMAW gun?
A. The pull type is used for soft and small diameter wires.
B. A trigger switch controls the wire feed.
C. Different nozzles are available for different applications.
D. The push type gun is used with heavier diameter wires.
E. all of the above

__________ **18.** Which of the metallurgical benefits resulting from the high speed of the GMAW process is not true?
A. There is a narrower heat-affected zone.
B. There is less grain growth.
C. There is less heat transfer in the parent metal.
D. There is greatly reduced distortion.
E. none of the above

__________ **19.** What is the meaning of the combined welding symbols shown?

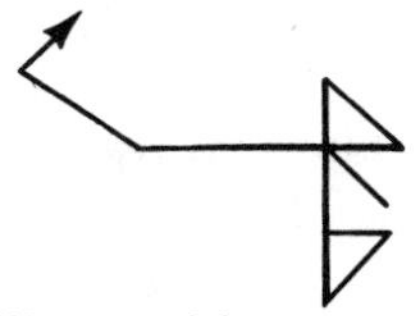

A. J-groove with fillet welds
B. Flare-groove with fillet welds
C. Bevel-groove with fillet welds both sides
D. V-groove with fillet weld arrow side
E. none of the above

Matching

__________ **1.** __________ results directly from atmospheric contamination.

__________ **2.** __________ is caused by excessive heat in the weld zone.

__________ **3.** __________ occurs when the arc does not melt the base completely, causing molten metal to flow outside the weld area.

__________ **4.** __________ results from a lack of heat input.

__________ **5.** __________ occurs when shielding gas is removed before the weld metal solidifies.

A. Cold lap
B. Surface porosity
C. Crater porosity
D. Insufficient penetration
E. Burn-through

__________ **6.** __________ occurs when welding current is low or below transition current.

__________ **7.** __________ is most practical for welding with wire .045″ or less on thinner sections.

__________ **8.** __________ requires a high current density and is practical for heavy gauge metal.

A. Spray transfer
B. Globular transfer
C. Short-circuit transfer

__________ **9.** __________ results in inadequate shielding.

__________ **10.** __________ is when the gun is pointing toward the end of the weld.

__________ **11.** __________ is 90° when welding flat position beads.

__________ **12.** __________ controls welding amperage.

__________ **13.** __________ is when the gun is pointing towards the beginning of the weld.

A. Drag angle
B. Push angle
C. Work angle
D. Gas drift
E. Wire feed

______ **14.** The ______ transfer permits welding thin metals in all positions.
______ **15.** ______ is not recommended for the GMAW process.
______ **16.** The ______ is the force that squeezes the droplet of metal from the wire.
______ **17.** The ______ extends from the minimum value where the heat melts the electrode to the point where high current induces spray transfer.
______ **18.** ______ is when the electrode is positive and the work is negative.

A. transition current
B. short-arc
C. DCRP
D. DCSP
E. electromagnetic pinch

______ **19.** A ______ delivers wire, shielding gas, and welding current to the weld area.
______ **20.** The ______ prevents oxygen, nitrogen, and hydrogen from contaminating the weld.
______ **21.** The ______ is usually a rectifier type welding machine for GMAW.
______ **22.** A ______ drives the electrode from the wire spool to the gun and arc.
______ **23.** The ______ is the same as electrode extension.

A. power supply
B. wire feeder
C. welding gun
D. shielding gas
E. wire stickout

Identify the parts of the GMAW process shown.

______ **24.** Average arc length
______ **25.** Wire
______ **26.** Nozzle-to-work distance
______ **27.** Actual stick-out
______ **28.** Gas nozzle

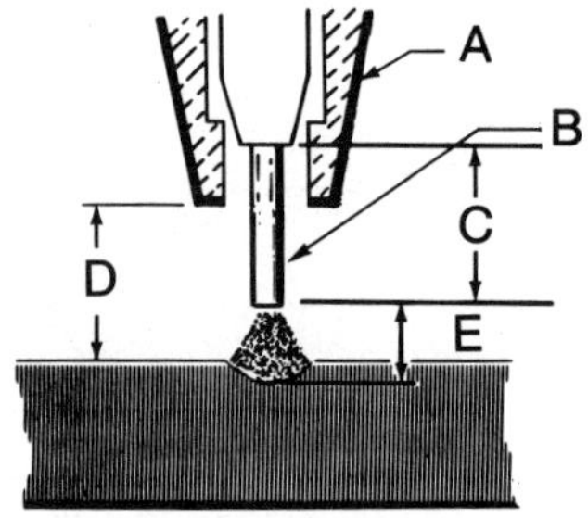

Identify the parts of the spray type of metal transfer shown.

______ **29.** Spray
______ **30.** Arc
______ **31.** Electrode
______ **32.** Weld
______ **33.** Bright inner cone

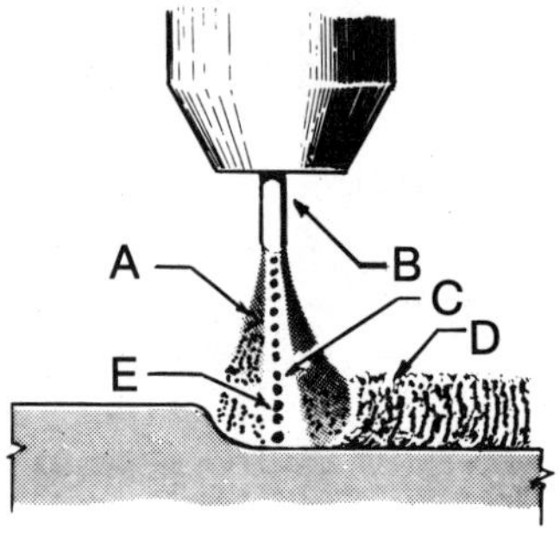

REVIEW QUESTIONS 26

Related Gas Metal-Arc Welding Processes

True-False

T F 1. Tubular wires are used when flux-cored arc welding.

T F 2. Pulsed-arc welding is used frequently for out-of-position welding.

T F 3. Globular transfer occurs in the peak current ranges of the pulsed-arc process.

T F 4. In pulsed-arc welding, the heat input range bridges the gap between the heat input ranges available from the spray and short-circuit arc processes.

T F 5. The pulse peak voltage is determined by the type and diameter of the electrode.

T F 6. The EXXT-2 flux-cored electrode does not require a shielding gas.

T F 7. Electrode extension should be approximately 1″ when running beads on mild steel using the FCAW process.

T F 8. The wire feeder must be adjusted to obtain the correct amperage.

T F 9. Voltage for flux-cored arc welding should be approximately 38–45 volts.

T F 10. The amperage is increased when welding in the flat position compared to the vertical position when flux-cored arc welding.

Multiple Choice

______ 1. Which of the following is not true about the submerged arc welding process?
- A. It is most often used for thick metals.
- B. Carbon dioxide is commonly used as a shielding gas.
- C. The weld can be applied by a semiautomatic or automatic process.
- D. Granular flux is used to protect the metal from atmospheric contamination.
- E. none of the above

______ 2. The flux-cored arc welding process most commonly uses ______ current.
- A. DCSP
- B. DCRP
- C. alternating
- D. alternating high frequency
- E. none of the above

______ 3. Which of the following is true about the buried-arc CO_2 welding process?
- A. Standard GMAW equipment is used.
- B. The metal transfer is globular.
- C. It allows deep penetration with fast deposition.
- D. Carbon dioxide is used as a shielding gas.
- E. all of the above

KNOW YOUR WELDING SYMBOLS

Welding Process

SMAW Indicates Shielded Metal-Arc Welding process to be used.

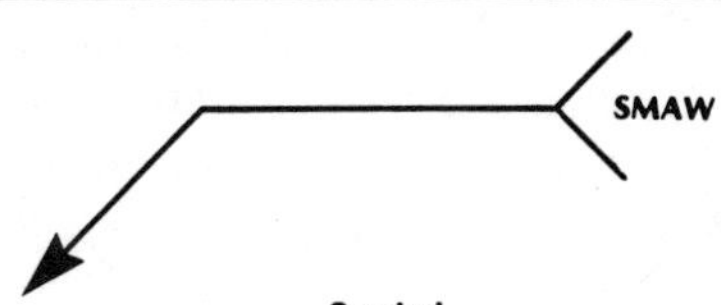

Desired Weld

Symbol

______ **4.** ______ is the shielding gas most commonly used for flux-cored arc welding.
A. Argon
B. Oxygen
C. Helium
D. Nitrogen
E. none of the above

______ **5.** The gas metal-arc welding—pulsed arc process ______.
A. has its peak current in the globular-transfer range
B. transfers metal by the short-circuit process only
C. produces a great amount of spatter
D. has its peak current in the spray transfer current range
E. none of the above

______ **6.** The AWS classification ______ carbon steel flux-cored electrode is designed to be used with DCEP current with CO_2 as a shielding gas.
A. EXXT-1
B. EXXT-3
C. EXXT-6
D. EXXT-10
E. all of the above

______ **7.** What is the meaning of the information in the reference tail shown?

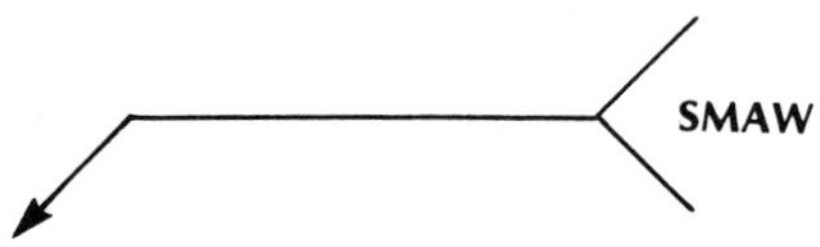

A. Spot weld
B. Submerged-arc welding
C. Spray arc
D. Shielded metal-arc welding
E. none of the above

Matching

______ **1.** ______ uses gaseous shield from flux to shield the weld area.

______ **2.** ______ uses a granular flux to completely cover the electric arc.

______ **3.** ______ uses current that is changed back and forth from the globular-transfer range to the spray-transfer range.

______ **4.** ______ uses a wire electrode at the level of or below the surface of the work.

______ **5.** ______ uses a shielding gas with a tubular wire.

A. Buried-arc CO_2 welding
B. Gas metal-arc welding—pulsed-arc
C. Flux-cored arc welding
D. Self-shielded flux-cored arc welding
E. Submerged-arc welding

Identify the parts of the FCAW-S process shown.

______ **6.** Base metal
______ **7.** Weld
______ **8.** Gaseous shield
______ **9.** Tubular electrode
______ **10.** Powdered flux

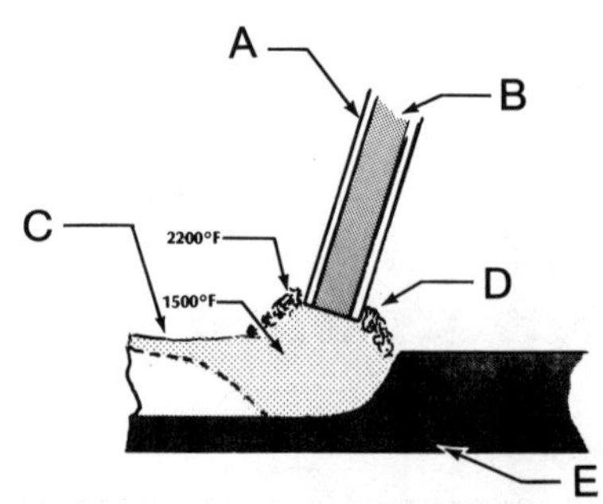

REVIEW QUESTIONS 27

Brazing

True-False

T F **1.** Braze welding should be done in the flat position.

T F **2.** The lowest effective heat is recommended when brazing to prevent any change in the mechanical properties of the base metal.

T F **3.** Surface oxides on the metal to be brazed can be eliminated by sanding, grinding, filing, blasting, or wire brushing.

T F **4.** Some brazing rods are coated with a flux.

T F **5.** The flux residue should be removed from the metal after brazing to prevent corrosion of the metal.

T F **6.** Brazing may require a melting temperature of 5,000°F.

T F **7.** The oxygen-argon flame produces the required heat necessary for brazing aluminum alloys.

T F **8.** A brazing filler metal must be completely molten before it flows into a joint.

T F **9.** A multiflame gas torch is one source of heat that can be used for brazing.

T F **10.** Remove any grease, dirt, or oil from the base metal before brazing.

T F **11.** The base metal should be melted slightly before adding any filler metal.

T F **12.** Brazing is done at temperatures above 800°F and below the melting temperature of the base metal.

T F **13.** A correctly deposited brazed bead shows uniform and consistent ripples.

T F **14.** When braze welding, no torch movement is necessary.

T F **15.** Soldering, unlike brazing, does not require flux.

T F **16.** When mixing acid and water, always pour the acid into the water to prevent undiluted acid from splashing out.

T F **17.** Sal-ammoniac is used to clean the point of the soldering copper.

T F **18.** When braze welding, preheat the base metal to a bright red.

Multiple Choice

__________ **1.** __________ is/are required to achieve sound brazed joints.

A. Correct joint design
B. Clean surfaces
C. Correct filler metals
D. Correct fluxes
E. all of the above

KNOW YOUR WELDING SYMBOLS

Weld-All-Around

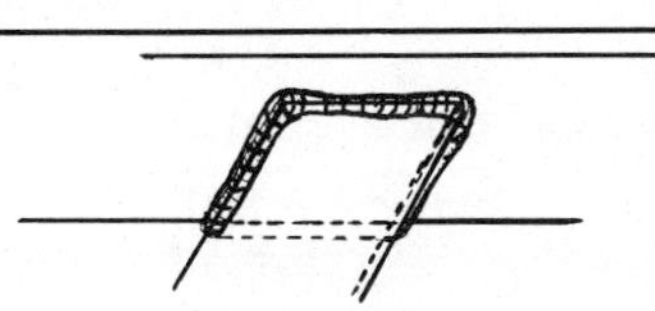

Desired Weld

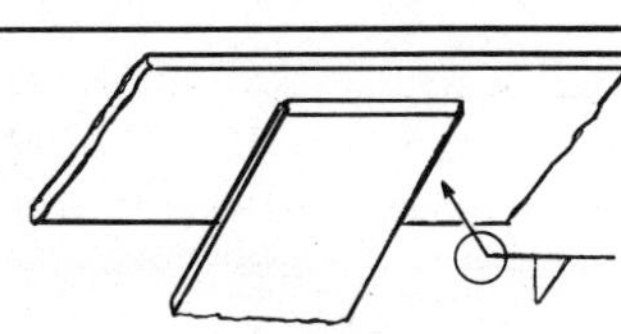

Symbol

______ **2.** Which of the following is not a consideration when determining the joint design of a brazed joint?
A. The joint should have adequate clearance to allow for capillary action.
B. The design of the joint is based on the adhesive qualities of the filler metal.
C. The lap joint offers the greatest strength.
D. The surfaces of the joint should be tight against each other for maximum strength.
E. all of the above

______ **3.** Clamps or jigs are used to ______________.
A. position the filler metal
B. clean the pieces to be brazed
C. allow flux to be removed easily
D. keep the pieces in alignment
E. all of the above

______ **4.** When braze welding, which of the following is recommended?
A. Work with a neutral flame.
B. Avoid welding metal that will be subjected to a high temperature later.
C. Bevel the edges of thick sections.
D. Do not weld a surface that has oil or grease on it.
E. all of the above

______ **5.** Which of the following is not an advantage of brazing as a form of joining metals?
A. Brazing produces stronger bonds in joints than welding.
B. Lower bonding temperatures reduce the possibility of a change in mechanical properties of the base metal.
C. Brazing can be used to join dissimilar metals.
D. The filler metal can be drawn into the joint by capillary action.
E. none of the above

______ **6.** What is the meaning of the welding symbol shown?

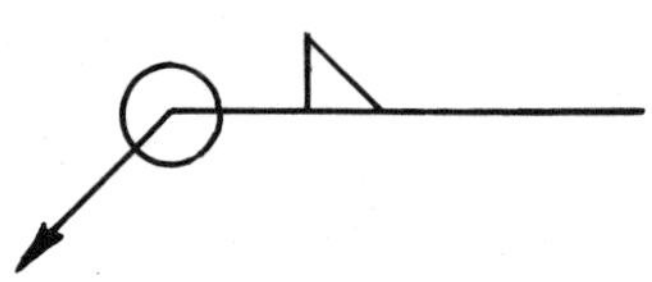

A. Weld around fillet arrow side
B. Weld around bevel-groove arrow side
C. Weld around V-groove arrow side
D. Weld around fillet other side
E. none of the above

Matching

______ **1.** ______ uses a filler metal composed of lead and tin.

______ **2.** ______ uses a filler metal that is not distributed by capillary action.

______ **3.** ______ uses silver as a filler metal.

______ **4.** ______ uses two basic joints—the butt and lap.

A. Brazing
B. Braze welding
C. Soft soldering
D. Hard soldering

______	5.	In the ______ brazing process, parts to be brazed are placed on trays.	A. resistance
______	6.	In the ______ brazing process, heat is generated in the same manner as it is in spot welding.	B. manual
______	7.	In the ______ brazing process, heat occurs by the resistance of the object to current flow through a coil.	C. dip
______	8.	In the ______ brazing process, parts to be brazed are immersed in molten brazing metal.	D. induction
______	9.	In the ______ brazing process, a gas torch is used to apply the required heat.	E. furnace

______	10.	______ inhibits the formation of oxides during the brazing process.	A. Solidus
______	11.	______ includes removing dirt, oil, grease, and oxides.	B. Liquidus
______	12.	______ is the highest temperature of the base metal reached in a solid state.	C. Flux
______	13.	______ is the melting temperature of the filler metal.	D. Surface preparation
______	14.	______ is available in wire, rod, strip, and powder form.	E. Filler metal

______	15.	______ involves applying a thin coating of solder.	A. Sweat soldering
______	16.	______ is the most common noncorrosive flux.	B. Seam soldering
______	17.	______ joins two pieces without any solder being visible.	C. Tinning
______	18.	______ involves running a layer of solder along the outside edge of the joint.	D. Rosin
______	19.	A ______ is heated inside a furnace to hold heat to melt the solder into the joint.	E. Soldering copper

		AWS Classification of Brazing Filler Metals	Types of Metals to be Brazed
______	20.	Copper, copper alloys	A. BAlSi (aluminum-silicon)
______	21.	Magnesium, magnesium alloys	B. BCuZn (copper-zinc)
______	22.	Stainless steels, carbon steels, low alloy steels, copper	C. BMg (magnesium)
______	23.	Ferrous and nonferrous metals	D. BCuP (copper-phosphorus)
______	24.	Aluminum, aluminum alloys	E. BNi (nickel)

		Approximate Melting Temperatures of Soft Solders	
______	25.	70% tin and 30% lead	A. 370°F
______	26.	50% tin and 50% lead	B. 471°F
______	27.	5% tin and 95% lead	C. 590°F

Identify the commonly used solder joints shown.

__________ **28.** Lap seam

__________ **29.** Single seam

__________ **30.** Joggle seam

__________ **31.** Grooved seam

__________ **32.** Double seam

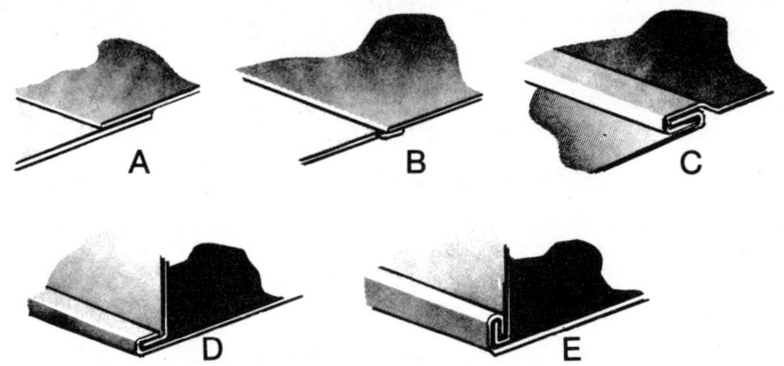

Identify the heating devices used for soldering shown.

__________ **33.** Electric soldering iron

__________ **34.** Soldering gun

__________ **35.** Soldering copper

__________ **36.** Soldering pencil

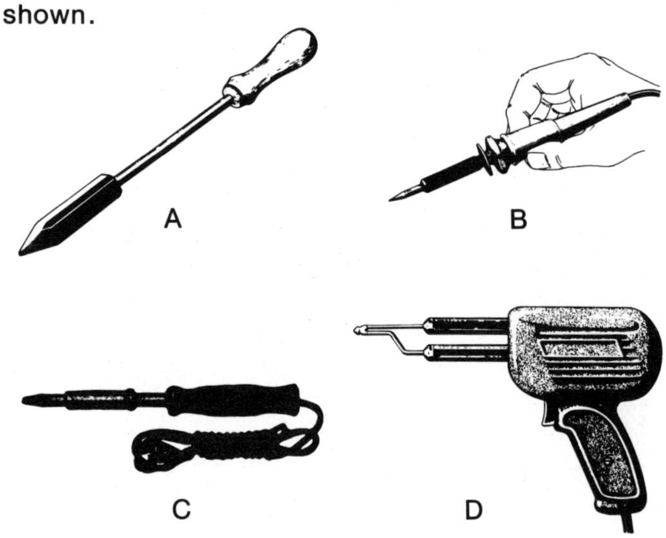

REVIEW QUESTIONS 28

Surfacing

True-False

T F 1. The Rockwell Hardness tester may be used to measure the hardness of a metal.

T F 2. Hardfacing using the gas metal-arc process allows faster application of overlays.

T F 3. The submerged-arc process is used when hardfacing parts where heavy deposits are required.

T F 4. Plasma-arc surfacing is a mechanized tungsten-arc process that uses a metal powder as surfacing material.

T F 5. When using a metallizing gun, each coating deposited is a layer .010″ thick.

T F 6. In the electric-arc metallizing process, the arc reaches approximately 7,000°F.

T F 7. The type of electrode used for hardfacing depends on the requirement of the weld metal applied.

T F 8. Stainless steel electrodes are often used to deposit base layers for other hardfacing electrodes.

T F 9. A slightly reducing flame is recommended when hardfacing with the oxyacetylene technique.

T F 10. All slag between layers (passes) should be removed when hardfacing with the shielded metal-arc process.

T F 11. Hardsurfacing deposits less metal per application compared to metallizing.

Multiple Choice

__________ 1. The high carbon electrodes used for hardfacing deposit weld metal that __________.

A. can be heat treated to produce harder surfaces
B. are applied with high amperage
C. can be hardened by cooling the part slowly
D. is resistant to severe abrasion
E. all of the above

__________ 2. The __________ motion is often used to prevent burn-through on thin sections.

A. crescent
B. whipping
C. drag
D. all of the above
E. none of the above

__________ 3. When hardfacing with the plasma-arc process, __________.

A. metal powder is carried from a hopper to the electrode holder in an argon gas stream
B. an AC power source is used
C. carbon dioxide is used to form the plasma
D. the hardfacing powder is carried to the arc with compressed air
E. none of the above

KNOW YOUR WELDING SYMBOLS

Scarf for Brazed Joint

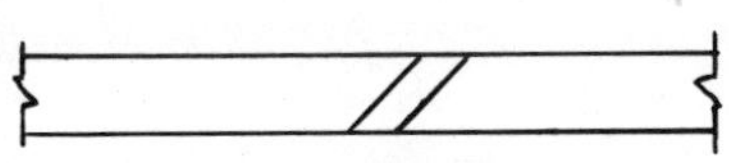

Desired Braze

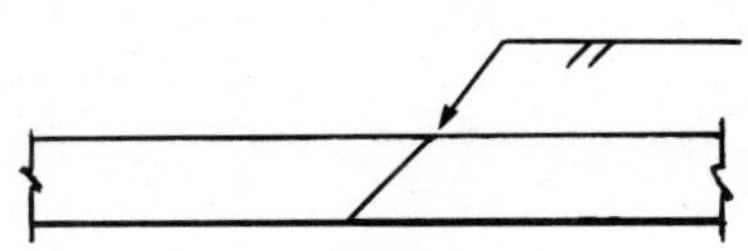

Symbol

_______ **4.** Metallizing _______________.
A. is a process of depositing fine semimolten metal particles onto the surface of a metal
B. is used to restore worn surfaces
C. is performed after the surface is thoroughly cleaned and roughed
D. can use a wire spray gun
E. all of the above

_______ **5.** To prevent weld hardening in medium and high-carbon steels, _______.
A. a high welding temperature is required
B. preheating and postweld heat treatment to the correct temperature is required
C. cooling should be rapid to prevent changes in the characteristics of the metal
D. all of the above
E. none of the above

_______ **6.** _______ is an element found in hardfacing materials.
A. Chromium
B. Tungsten
C. Silicon
D. Manganese
E. all of the above

_______ **7.** Which of the following is not true when hardfacing using the shielded metal-arc process?
A. Arrange the work for welding in the flat position.
B. A high amperage should be used to obtain maximum penetration.
C. A medium long arc should be maintained.
D. The slag should be removed between passes (additional layers).
E. all of the above

_______ **8.** Which of the following is not a recommended procedure when using the oxyacetylene hardfacing technique?
A. Use high-carbon filler rods.
B. Use a forehand welding technique.
C. Preheating is not recommended.
D. A slight weaving motion is used.
E. all of the above

_______ **9.** Severe abrasion-resistant electrodes for hardfacing _______________.
A. are not suitable for impact wear
B. are of the tungsten carbide and chromium carbide types
C. deposit a very hard abrasive-resistant material
D. come in either coated tubular form or as regular coated cast alloy
E. all of the above

_______ **10.** What is the meaning of the welding symbol shown?

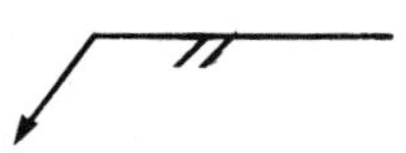

A. Butt joint arrow side
B. Butt joint other side
C. Scarf for brazed joint other side
D. Scarf for brazed joint arrow side
E. none of the above

REVIEW QUESTIONS 29

Pipe Welding

True-False

T F **1.** Most pipe welding is done with the shielded metal-arc process.

T F **2.** Pipe welders are certified for specific welding tasks.

T F **3.** Thick-wall pipe weld joints are usually closed butt joints.

T F **4.** An advantage of using a gas shielded-arc process is that there is no slag to be removed.

T F **5.** Consumable insert rings are used to maintain a consistent root opening around the pipe joint.

T F **6.** Tack welds should not penetrate the root of the groove.

T F **7.** E-6010 and E-6011 electrodes are commonly used for pipe welding.

T F **8.** A properly deposited root bead penetrates the root with a slight crown on the inside of the pipe and should not exceed ¼″.

T F **9.** The hot pass is used to give the weld joint a neat appearance.

T F **10.** Filler passes are usually made with a side-to-side electrode motion.

T F **11.** To restart a weld, the arc is struck approximately ½″ back of the bead and then moved forward with a long arc length.

T F **12.** Large diameter pipe is often joined using automatic gas metal-arc welding equipment.

T F **13.** A weaving motion is used when depositing the root pass.

T F **14.** Special pipe clamps are used to keep the pipes in alignment while tack welds are made.

T F **15.** Backing rings used in pipe welding to prevent spatter and slag from entering the pipe at the joint.

T F **16.** Whipping the electrode permits better control of the weld puddle.

Multiple Choice

__________ **1.** Which of the following is true about pipe welding?
- A. Pipe welding is recognized as a trade in itself.
- B. Certification standards vary for different pipe welding jobs.
- C. It is the easiest and simplest method of joining sections of pipe.
- D. Most pipe welding is done with the shielded metal-arc process.
- E. all of the above

__________ **2.** The roll welding method of pipe welding is performed by welding in the ______________________________ position.
- A. flat
- B. horizontal
- C. vertical
- D. overhead
- E. all of the above

KNOW YOUR WELDING SYMBOLS

Field Weld

Weld to be made in field

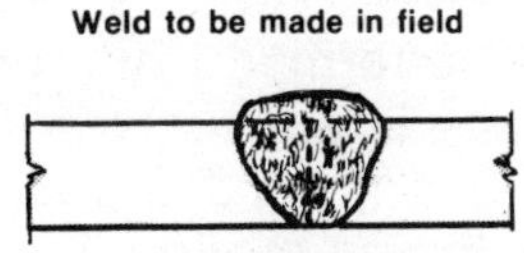

Desired Weld

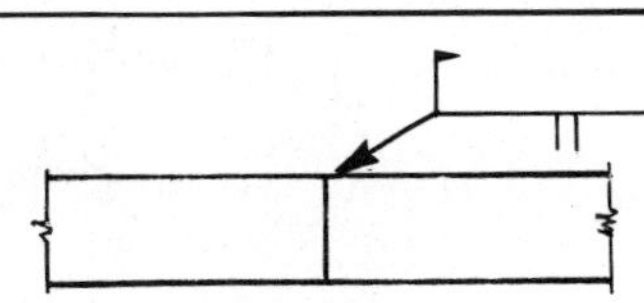

Symbol

_______ 3. The stove pipe or position method of pipe welding is performed by _______.
A. welding on the pipe in flat, horizontal, vertical, and overhead positions
B. lining up each section length by length before welding
C. keeping the pipe in a stationary position
D. not rotating the pipe
E. all of the above

_______ 4. A _______ is made by the penetration of the root bead (pass) and must be completely filled when restarting the root pass.
A. slag inclusion
B. porosity
C. slag
D. keyhole
E. none of the above

_______ 5. A properly deposited root bead should penetrate to the root and leave a solid bead below the surface with a slight crown not exceeding _______.
A. 1/2″
B. 1/4″
C. 1/16″
D. 5/16″
E. none of the above

_______ 6. Pipes having wall thicknesses of 1/8″ to 5/16″ are classified as _______.
A. rigid wall
B. thick wall
C. thin wall
D. backing ring pipe
E. all of the above

_______ 7. Small diameter pipes with wall thicknesses less than 1/8″ _______.
A. require a 75° groove included angle
B. are welded using the vertical up procedure
C. are beveled on both sides of the butt joint
D. are welded without any edge preparation
E. all of the above

_______ 8. Tack welds used in pipe welding _______.
A. are evenly spaced around the pipe
B. maintain the correct between-pipe spacing
C. are positioned in four locations on the pipe
D. are approximately 3/4″ long
E. all of the above

_______ 9. Which of the following does not apply to welding thin-wall pipe?
A. The vertical down technique is used.
B. The first pass deposited is the root pass.
C. The weld is started at the twelve o'clock position.
D. Heat input can be reduced by using a larger diameter electrode.
E. none of the above

_______ 10. Which of the following is not true about automatic gas metal-arc welding?
A. The bevel for automatic is different compared to conventional pipe welding.
B. All welds are made uphill.
C. The automatic gas metal-arc process is faster than stick welding.
D. The root bead is applied on the inside of the pipe.
E. none of the above

_______ 11. Which of the following refers to welding heavy-wall pipe?
A. Welding is performed from the 6:30 position upward.
B. The electrode is advanced with a travel angle of 5°–10°.
C. After the root pass is completed, one or more filler passes are deposited.
D. A slight whipping motion may be used to control the puddle.
E. all of the above

__________ **12.** What is the meaning of the welding symbol shown?

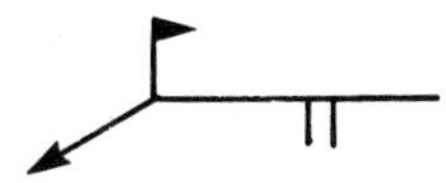

A. V-groove arrow side, weld made in field
B. V-groove other side, weld made in field
C. Butt joint arrow side, weld made in field
D. Bevel-groove arrow side
E. none of the above

Matching

__________ **1.** The function of the ________ is to burn out remaining particles of slag that may be in the groove.

__________ **2.** The ________ is made with a side-to-side weaving motion.

__________ **3.** The ________ provides a neat appearance and weld reinforcement.

__________ **4.** The success of the pipe weld depends on complete penetration of the ________.

A. stringer (root) bead
B. hot pass
C. filler pass
D. cap pass

Question 5–9 refer to Pipe Weld Joint Test Positions

__________ **5.** In the horizontal fixed position, the axis of the pipe is horizontal but the pipe is not turned or rolled during welding.

__________ **6.** This test position does not exist in pipe welding.

__________ **7.** The axis of the pipe is at a 45° angle.

__________ **8.** Welding is done in the flat position with the pipe rotating under the arc.

__________ **9.** Horizontal welding with the axis of the pipe in vertical position and the axis of the weld in horizontal position.

A. 1G
B. 2G
C. 3G and 4G
D. 5G
E. 6G

Identify the consumable insert rings shown.

__________ **10.** Class 1
__________ **11.** Class 2
__________ **12.** Class 4
__________ **13.** Class 5

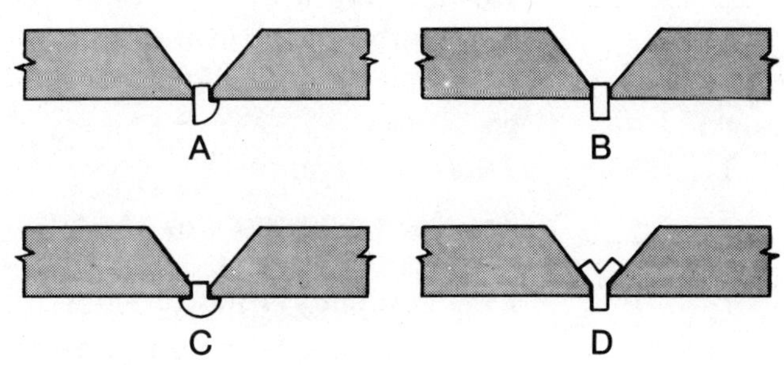

REVIEW QUESTIONS 30

Cutting Operations

True-False

T F **1.** Oxyacetylene flame cutting is one of the most popular methods used to cut metal.

T F **2.** Mapp, natural gas, propane, and acetylene can be used with oxygen to cut metal.

T F **3.** Cutting metal rapidly oxidizes the metal, thus reducing the material that is cut away into iron oxide (slag).

T F **4.** The cutting torch has five needle valves used to control the flame.

T F **5.** Preheat holes are used to blow the molten metal away from the cutting section.

T F **6.** The correct oxygen and acetylene pressure of a particular torch is determined by following the manufacturer's recommendations.

T F **7.** The size of the cutting tip used depends on the gas flow that is available.

T F **8.** To ignite the flame on a cutting torch, the acetylene needle valve must be opened first.

T F **9.** To flame cut round stock, the cut should be started approximately 90° from the top edge.

T F **10.** The preheat flame should be adjusted so that it is slightly carburizing.

T F **11.** If the cut is penetrating the thickness of the plate, a shower of sparks will be seen on the underside of the plate.

T F **12.** A bar clamped parallel to the cut can serve as a guide for a straight cut.

T F **13.** When cutting cast iron, the cutting operation should be stopped and started as necessary to complete the cut.

T F **14.** The plasma-arc cutting process can cut steel up to ten times faster than the conventional flame cutting process.

T F **15.** The tip of the electrode in a plasma-arc torch is located within the nozzle.

T F **16.** When plasma-arc cutting, the arc must be stopped at the power source when the cut is complete.

T F **17.** The air carbon-arc cutting process uses compressed oxygen with a carbon arc to blow molten metal away from the cutting zone.

T F **18.** Copper clad electrodes last longer than plain carbon electrodes used for air carbon-arc cutting.

T F **19.** The preheat flame should be slightly carburizing for cutting cast iron.

T F **20.** Direct current straight polarity is used for plasma-arc cutting.

T F **21.** The head of the cutting torch is tilted when making a bevel cut.

T F **22.** When performing any cutting operation, be sure there are no combustible materials nearby.

T F **23.** Washing is the process of removing metal from large areas.

KNOW YOUR WELDING SYMBOLS

Melt-Thru Dimension

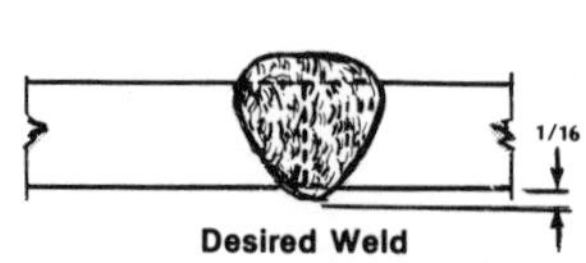

Desired Weld

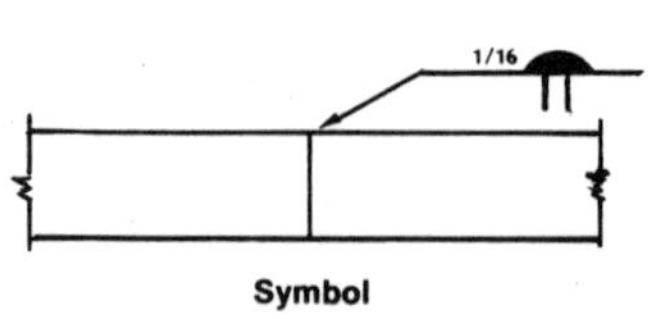

Symbol

Multiple Choice

_______ 1. The _______ control lever activates the flow of gas through the center hole of the cutting tip.
   A. acetylene
   B. preheat
   C. carbon arc
   D. oxygen
   E. none of the above

_______ 2. The air carbon-arc cutting process requires compressed air in the range of _______ psi.
   A. 100–200
   B. 10–20
   C. 40–80
   D. 400–800
   E. none of the above

_______ 3. When flame cutting, the torch is held so that the inner cone is approximately _______″ above the metal.
   A. $^1/_4$
   B. $^1/_2$
   C. $^1/_8$
   D. $^1/_{16}$
   E. none of the above

_______ 4. The pressure of gases needed to cut using the oxyacetylene flame is determined by the _______.
   A. length of the metal being cut
   B. torch tip number
   C. type of metal being cut
   D. properties of the metal being cut
   E. none of the above

_______ 5. A number _______ cutting tip is recommended for cutting metal ½″ thick.
   A. 1
   B. 2
   C. 3
   D. 0
   E. none of the above

_______ 6. _______ is the most efficient gas to use in the plasma-arc cutting process.
   A. Oxygen
   B. Hydrogen
   C. Nitrogen
   D. Air
   E. none of the above

_______ 7. The air jet orifices must be positioned _______ the electrode when using the air carbon-arc cutting torch.
   A. above
   B. under
   C. toward
   D. over
   E. all of the above

_______ 8. The cutting flame and the preheat flame of the cutting torch should be _______.
   A. oxidizing
   B. neutral
   C. carburizing
   D. reducing
   E. none of the above

_____ 9. A _____ movement of the cutting torch is continued throughout the length of the cut on cast iron.
   A. swinging
   B. whipping
   C. up-and-down
   D. in-and-out
   E. none of the above

_____ 10. To make a mechanized cut using the plasma-arc cutting process, locate the center of the torch approximately _____" above the surface of the plate to be cut.
   A. 1/2
   B. 3/4
   C. 7/8
   D. 1/4
   E. all of the above

_____ 11. What is the meaning of the dimension shown on the welding symbol?

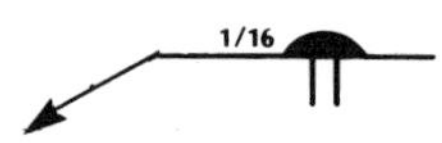

   A. Scarf joint the 1/16" root opening
   B. Butt joint with 1/16" root opening
   C. Surface weld with 1/16" throat
   D. Butt joint with 1/16" melt-thru
   E. none of the above

Matching

Identify the parts of the air carbon-arc cutting torch shown.

_____ 1. Air and power supply

_____ 2. Air jet orifices

_____ 3. Starting button

_____ 4. Electrode

_____ 5. Electrode release

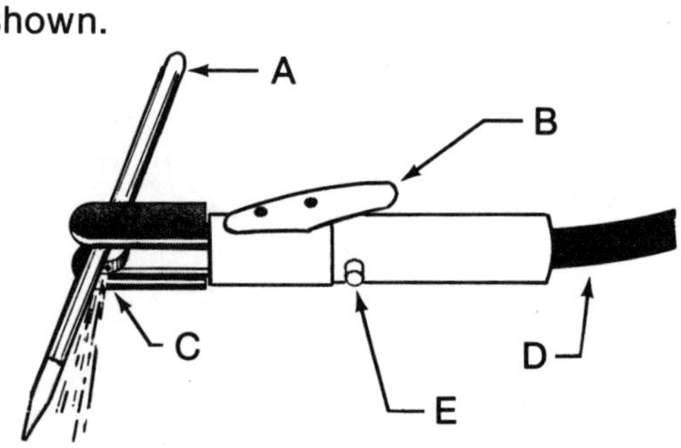

Identify the parts of the plasma-arc cutting torch shown.

_____ 6. Plasma arc

_____ 7. Gas

_____ 8. Nozzle

_____ 9. Tungsten cathode

_____ 10. Kerf

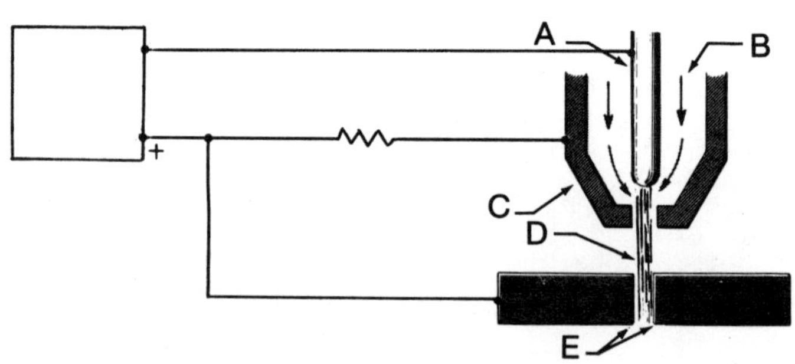

REVIEW QUESTIONS 31

Production Welding

True-False

T F **1.** Amperage in the gas tungsten-arc spot welding process should be adjusted as necessary to allow for the thickness of the metal being welded.

T F **2.** In inertia welding, the pieces are fused with heat that is generated by friction from a piece rotating against a stationary piece.

T F **3.** Resistance welding is often used in production welding operations.

T F **4.** Spot welding is a type of resistance welding.

T F **5.** Spot welders require more than one electrode to perform the welding operation.

T F **6.** Current flow in pulsation spot welding is interrupted by precise electronic control.

T F **7.** Pressure must be applied from both sides of the weld in gas tungsten-arc spot welding.

T F **8.** Both weld pieces must be in contact with each other in gas tungsten-arc spot welding.

T F **9.** Fusion in electron beam welding is achieved with a high-power density beam focused on the area to be joined.

T F **10.** The beam-in-air electron beam welding process requires a vacuum chamber to prevent contamination from entering the weld area.

T F **11.** Inertia welding is sometimes called friction welding.

T F **12.** The heat affect zone in inertia welding is narrow compared to that in other welding processes.

T F **13.** Laser welding utilizes a highly concentrated beam to generate a power intensity of more than one billion watts.

T F **14.** Direct current straight polarity is used in the plasma-arc welding process.

T F **15.** Argon is often used as plasma gas in the plasma-arc welding process.

T F **16.** The ultrasonic welding process uses vibratory energy to fuse the weld pieces together.

T F **17.** Roller spot welding can be used to obtain a continuous seam in the resistance welding process.

T F **18.** Electro-gas welding is used for multiple-pass welds on horizontal joints.

T F **19.** The beam-in-air electron beam welding process produces welds with characteristics similar to those in the gas tungsten-arc welding process.

T F **20.** Projection welding is a form of resistance welding that uses raised points which become part of the weld area.

T F **21.** Ultrasonic welding uses heat to fuse metals together.

T F **22.** Stud welding is commonly used for fusing light gauge metals.

T F **23.** Spot welding is used to join metals ½″ or greater in thickness.

KNOW YOUR WELDING SYMBOLS

Surface Weld Dimension

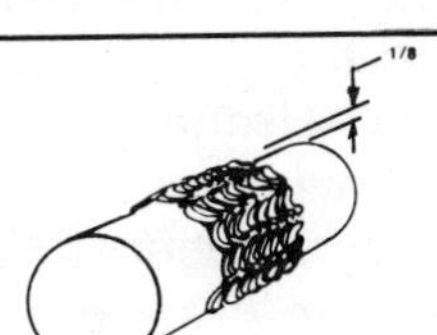

Desired Weld

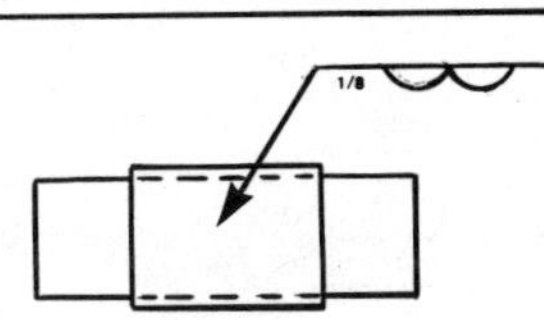

Symbol

Multiple Choice

_______ 1. In resistance welding, _______________.
- A. fusion takes place when pressure is applied to metal in a plastic state
- B. a gas supported flame provides fusion heat
- C. electrodes do not conduct electricity to the work
- D. thick metals can be spot welded
- E. all of the above

_______ 2. The jaws of spot welders may be operated _______.
- A. hydraulically
- B. manually
- C. pneumatically
- D. all of the above
- E. none of the above

_______ 3. Spot welders can use _______ current.
- A. alternating and direct
- B. high voltage
- C. low amperage
- D. static
- E. all of the above

_______ 4. Gas tungsten-arc spot welding _______ compared to conventional resistance spot welding.
- A. provides deeper, more localized penetration
- B. does not require accessibility to both sides of the joint
- C. requires a shielding gas
- D. requires a tungsten electrode
- E. all of the above

_______ 5. When gas tungsten-arc spot welding, using _______ as a shielding gas provides deeper penetration.
- A. argon
- B. oxygen
- C. carbon dioxide
- D. helium
- E. none of the above

_______ 6. _______ welding utilizes the Graham and Nelson method.
- A. Gas metal-arc
- B. Beam
- C. Pulsation
- D. Stud
- E. none of the above

_______ 7. Plasma-arc welding generates the heat for fusion by an electric arc that has been intensified by the injection of _______ into the arc stream.
- A. voltage
- B. ions
- C. gas
- D. tungsten
- E. none of the above

_______ 8. Equipment necessary for ultrasonic welding includes two components: the _______ and the _______.
- A. power source, electrode
- B. shielding gas, flowmeter
- C. power source, transducer
- D. vibration generator, grounding mechanism
- E. all of the above

______ **9.** What is the meaning of the dimension shown on the welding symbol?

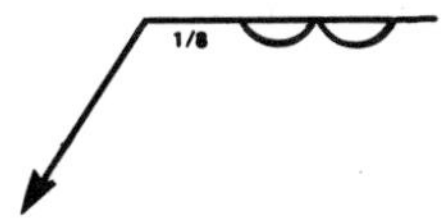

A. Surface weld $^1/_8$" in length
B. Melt-thru $^1/_8$"
C. Surface weld $^1/_8$" in thickness
D. Surface weld
E. none of the above

Matching

Identify the four principal elements of a standard resistance welder.

______ **1.** Electrode
______ **2.** Timing control
______ **3.** Frame
______ **4.** Electrical circuit

A. reduces voltage and increases amperage to provide necessary heat
B. regulates volume and length of current
C. is the main body of the machine
D. is the mechanism for making and holding contact at the weld area

Identify the principle types of resistance welding.

______ **5.** Spot welding uses ______.
______ **6.** Seam welding uses ______.
______ **7.** Projection welding uses ______.
______ **8.** Flash welding uses ______.
______ **9.** Butt welding uses ______.

A. raised sections of the weld pieces to make contact, melt, and become part of the weld area
B. constant pressure applied during the heating process
C. two metal pieces moved together until an arc is established
D. applied pressure on the metal pieces to be joined, and electricity is sent from one electrode to another
E. roller type electrodes to produce a continuous weld seam

Identify the parts of the plasma-arc process shown.

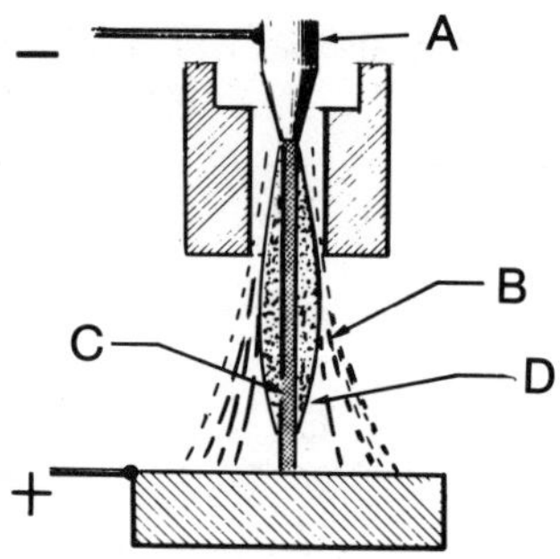

_______ **10.** Inner (hot) sheath

_______ **11.** Outer (cool) sheath

_______ **12.** Tungsten electrode

_______ **13.** Arc core

Identify the parts of the laser welding process shown.

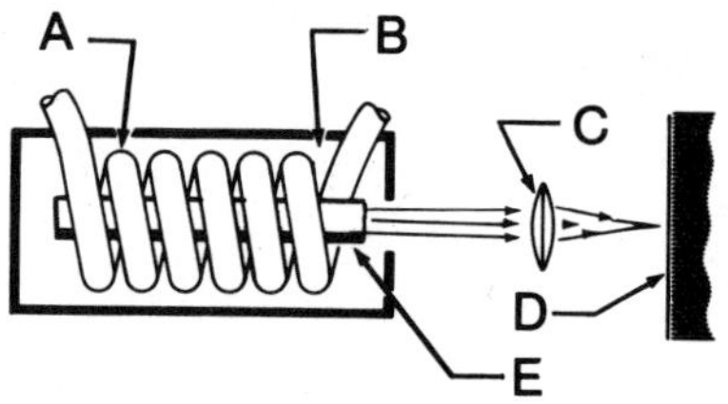

_______ **14.** **Lens**

_______ **15.** **Optical cavity**

_______ **16.** **Pumping source**

_______ **17.** **Target**

_______ **18.** **Laser crystal**

Identify the adhesive used in adhesive bonding.

_______ **19.** Epoxy adhesive

_______ **20.** Hot melt adhesive

_______ **21.** Cyanoacrylate adhesive

_______ **22.** Solvent-base adhesive

_______ **23.** Silicone adhesive

A. Cures instantly by reacting to trace surface moisture
B. Two-part adhesive that cures when resin and hardener are combined
C. Has high temperature resistance and excellent sealing characteristics
D. Thermoplastic material applied in molten state
E. Commonly used as contact cement for bonding large surface areas

REVIEW QUESTIONS 32

Plastic Welding

True-False

T F **1.** Applying too much heat to the weld area results in a charred and discolored weld.

T F **2.** For best results, triangular-shaped filler rods are used for fillet welds.

T F **3.** Ventilation is not required for plastic welding.

T F **4.** A fanning motion is used to distribute uniform heat over the rod and the edges of the joint.

T F **5.** Plastic welding requires that the welder wear a welding helmet for protection from harmful rays.

T F **6.** The filler rod used should match the properties of the plastic being welded.

T F **7.** Thermosetting plastics can be softened repeatedly for different manufacturing processes.

T F **8.** Plastic softens when heated, unlike metal, which melts when heated.

T F **9.** The same joints used in welding metals are used for welding plastic.

T F **10.** Different welding tips are required for specific joint applications.

T F **11.** Nitrogen gas is best used when welding PVC plastics.

T F **12.** Plexiglass requires a higher welding temperature than PVC plastic.

T F **13.** Filler rods come in flat, round, and triangular shapes.

T F **14.** A tacker tip is used to fuse the pieces together prior to welding.

T F **15.** A special torch is used to apply filler metal to the weld piece in friction welding.

T F **16.** Friction welding is sometimes called spin welding.

T F **17.** In induction welding, current flows through a metal insert to generate heat in the weld area.

T F **18.** Consult the manufacturer to determine the correct air and gas pressure necessary for different welding applications.

T F **19.** A round filler rod with a special tip is used in high speed welding.

T F **20.** The end of a filler rod should be cut at an angle of approximately 60°.

Multiple Choice

________ **1.** ________ is used in hot gas welding.

A. Compressed air
B. Nitrogen gas
C. A gun containing an electrical heating unit
D. A different tip for each different welding task
E. all of the above

KNOW YOUR WELDING SYMBOLS

Flush Contour of Weld

Desired Weld

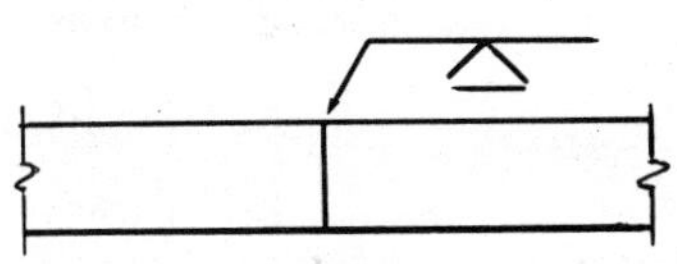

Symbol

________ **2.** The ________ plastic family will soften only once when exposed to heat.
A. thermoplastic
B. thermosetting
C. thermomolten
D. phenolic
E. none of the above

________ **3.** ________ will soften repeatedly whenever heat is applied.
A. Polyvinyl chlorides
B. Polyethylenes
C. Polypropylenes
D. all of the above
E. none of the above

________ **4.** Hot gas welding guns supply a welding temperature of approximately ______°F.
A. 1,200–1,400
B. 204–316
C. 400–925
D. 204–496
E. none of the above

________ **5.** ________ are not in the thermosetting plastic family.
A. Polyesters
B. Ureas
C. Acrylics
D. Epoxies
E. none of the above

________ **6.** All butt joints should be ________ to obtain a strong bond when welding.
A. square
B. closed square
C. beveled
D. square groove
E. all of the above

________ **7.** What is the meaning of the welding symbol shown?

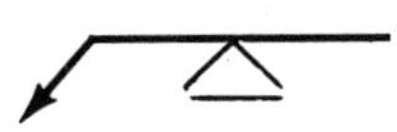

A. Bevel-groove other side, flush contour
B. V-groove arrow side, flush contour
C. Double bevel-groove other side, concave contour
D. Flange weld with convex contour
E. none of the above

Matching

Identify the parts of the high-speed welding process shown.

________ **1.** Base material

________ **2.** High-speed tool

________ **3.** Plasticized strip

________ **4.** Position of welder

________ **5.** Flowlines

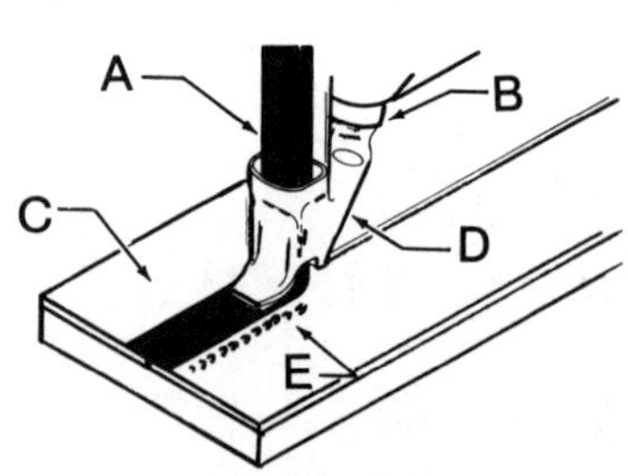

_______ **6.** Friction welding occurs when the _______.

_______ **7.** Heated-tool welding occurs when the _______.

_______ **8.** Induction welding occurs when the _______.

_______ **9.** Hot gas welding occurs when the _______.

A. edges to be joined are heated to fusing temperature and brought in contact

B. heat for fusing is caused by rubbing the surfaces of the parts to be joined

C. welding gun uses a heating element and compressed gas to soften weld pieces and filler rod

D. current flowing through a metal insert creates heat for fusion

REVIEW QUESTIONS 33

Robotics and Welding

True-False

T F 1. Robotic welding systems use basically the same welding equipment used for semiautomatic and machine applications.

T F 2. Two commonly used robots are the rectilinear and articulating robots.

T F 3. Welding robots are inexpensive machines that do not require regular maintenance.

T F 4. The welding process used determines the type of robot that should be used.

T F 5. A positioner is used to locate the part being welded in the position for the specific welding operation.

T F 6. The working volume of the robot is the amount of parts that can be welded in a 10-minute period.

T F 7. Robot welding systems commonly use the oxyacetylene welding process for joining metals.

T F 8. The robot controller coordinates movements between the robot and the positioner.

T F 9. Robot welders are used in industry for welds that are repeated rapidly in the manufacturing process.

T F 10. A five axes robot has five movable joints to facilitate movement in and around the weld area.

T F 11. The rectilinear type robot is used primarily for welding pieces larger than those welded with the articulating type robot.

T F 12. Injury is avoided by standing clear of any automatic welding machine.

T F 13. A fixed-position jig can be used to hold the work in a stationary position.

T F 14. Depending on the type of robot, axis movement can be accomplished hydraulically or electrically.

T F 15. A robot is sometimes called a manipulator.

Multiple Choice

_______ 1. A _______ maneuvers in and around the work to perform the welding task.
- A. positioner
- B. robot
- C. control programmer
- D. jig
- E. none of the above

_______ 2. The _______ translates and relays commands from the operator controls to the manipulator.
- A. operator controls
- B. robot
- C. control pendant
- D. robot controller
- E. none of the above

KNOW YOUR WELDING SYMBOLS

Convex Contour of Weld

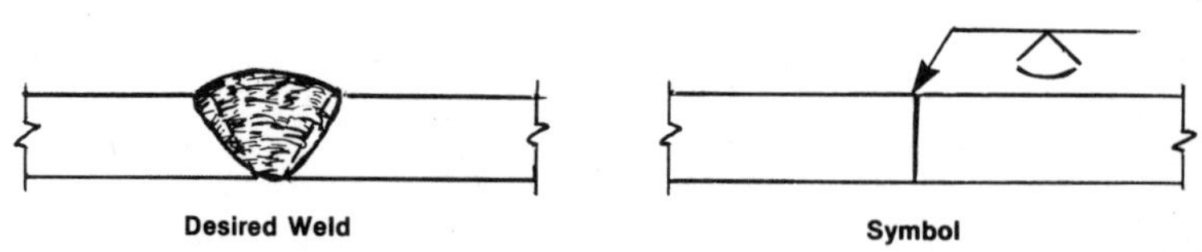

_______ **3.** The _______ is the same equipment used for machine and semiautomatic applications.

A. positioner
B. jig
C. operator controls
D. automatic welding equipment
E. all of the above

_______ **4.** The rectilinear-type robot _______.

A. is used to weld large pieces
B. has an extendable horizontal axis
C. can be used at two work stations on the horizontal axis
D. is used for tasks that require a larger working volume than an articulating robot
E. all of the above

_______ **5.** A control pendant is used for _______.

A. semiautomatic welding applications
B. on-site programming of the welding system
C. determining the working volume
D. all of the above
E. none of the above

_______ **6.** What is the meaning of the welding symbol shown?

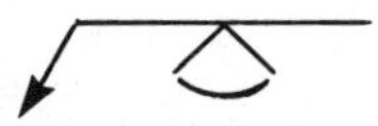

A. Bevel-groove other side, flush contour
B. V-groove arrow side, convex contour
C. Double V-groove, convex contour
D. V-groove other side, flush contour
E. none of the above

Matching

Identify the major components of the robot welding system shown.

_______ **1.** Robot
_______ **2.** Robot controller
_______ **3.** Positioner
_______ **4.** Automatic welding equipment
_______ **5.** Operator controls

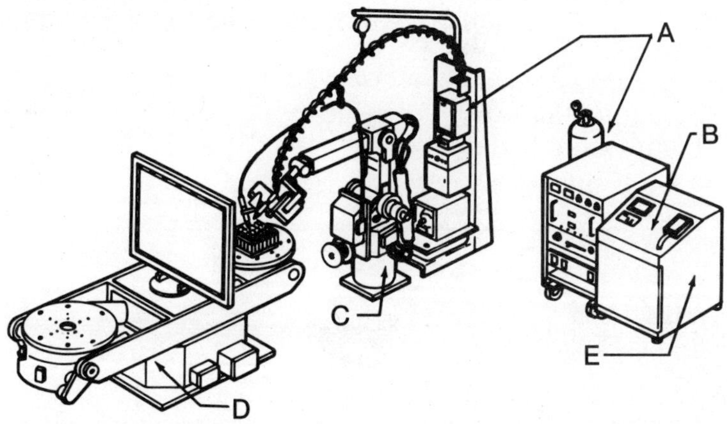

REVIEW QUESTIONS 34

Testing Welds

True-False

T F **1.** In tensile testing, the weld piece is pulled until it breaks.

T F **2.** Nondestructive testing deforms the weld piece.

T F **3.** Visual inspection is a form of nondestructive testing.

T F **4.** A guided bend test shows the degree of penetration achieved by the weld.

T F **5.** Internal weld defects can be checked with radiographic testing equipment.

T F **6.** A fillet weld is tested with the free bend testing technique.

T F **7.** Impact testing is used to test the ductility of a weld.

T F **8.** A Rockwell hardness tester determines relative hardness of the weld area compared with the base metal.

T F **9.** The dye penetrant inspection method locates defects in the weld by outlining surface defects.

T F **10.** A universal testing machine can be used to test the strength of a weld.

T F **11.** The percentage of elongation of the weld specimen is found by fitting the broken ends of the tested piece and measuring the new gauge length.

T F **12.** In the face bend test, the specimen is placed with the weld face down in the guided-bend jig.

T F **13.** The soundness and ductility of a weld can be ascertained by the nick-break test.

T F **14.** A black light can be used with the dye penetrant inspection method.

T F **15.** Two types of specimens used for impact testing are Charpy and Izod.

T F **16.** The Brinell test is used to check the tensile strength of a weld specimen.

T F **17.** The Rockwell hardness tester uses a variety of loads and indenters necessitating different scales.

T F **18.** When preparing a free bend weld specimen, the scratches produced by grinding should run parallel to the bend made.

T F **19.** Hydrochloric acid can be used in the dye penetrant inspection method.

T F **20.** When diluting acid, the acid should always be poured into water to prevent full strength acid from splashing out.

Multiple Choice

__________ **1.** The __________ method is a nondestructive testing procedure.
- A. guided-bend testing
- B. shear testing
- C. etch testing
- D. ultrasonic testing
- E. all of the above

KNOW YOUR WELDING SYMBOLS

Concave Contour of Weld

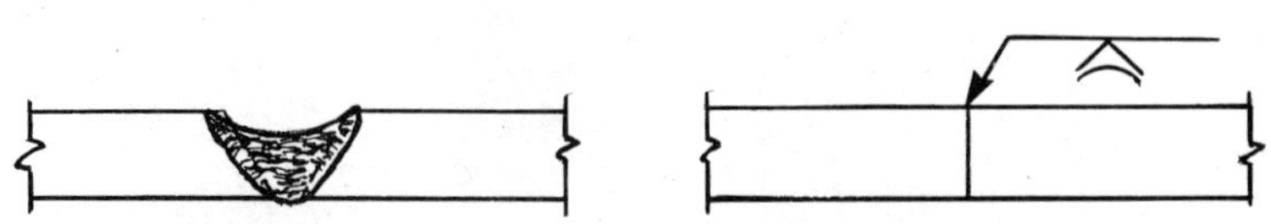

Desired Weld   Symbol

______ **2.** ______ is not a weld defect.

A. Slag inclusion
B. Overlap
C. Porosity
D. Tensile strength
E. none of the above

______ **3.** The ______ test is a common type of destructive testing.

A. etching
B. impact
C. shear
D. weld uniformity
E. all of the above

______ **4.** The guided-bend test requires ______ to check the degree of fusion and weld penetration.

A. a liquid dye
B. a fillet weld
C. ammonia persulfate
D. two specimens
E. gauge marks

______ **5.** Visual examination is used to ______.

A. locate internal weld defects
B. detect grain growth
C. check for overlaps
D. measure tensile strength
E. all of the above

______ **6.** What is the meaning of the welding symbol shown?

A. V-groove other side, concave contour
B. V-groove arrow side, convex contour
C. Bevel-groove, flush contour
D. Bevel-groove both sides
E. none of the above

Matching

Identify the various testing methods used to determine the quality of a weld.

______ **1.** In the ______ method, surface defects are located by dyes.

______ **2.** In ______, high-frequency vibrations are used to locate defects.

______ **3.** The ______ method uses electromagnetic and sensitized film.

______ **4.** The ______ method uses iron powder suspended in a liquid.

______ **5.** The ______ method produces a current in the test piece by induction.

A. dye penetrant inspection
B. magentic particle inspection
C. ultrasonic testing
D. radiographic inspection
E. eddy current testing

Identify the weld test specimens shown below.

_____________ **6.** Charpy
_____________ **7.** Izod
_____________ **8.** Nick-break
_____________ **9.** Tensile specimen for butt weld

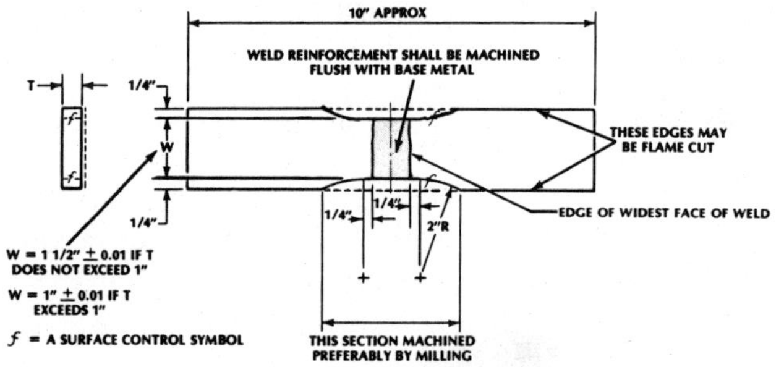

A

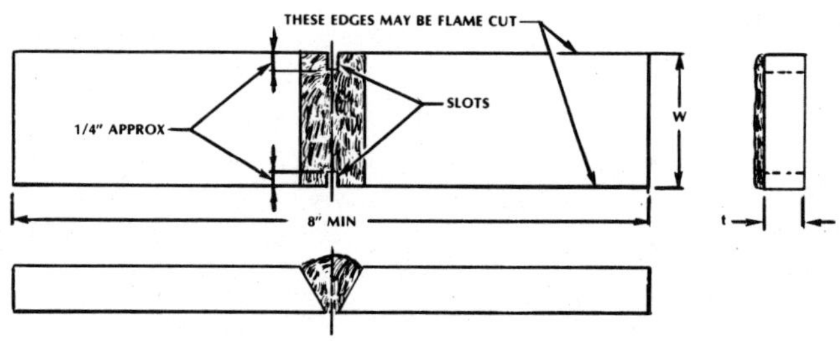

B

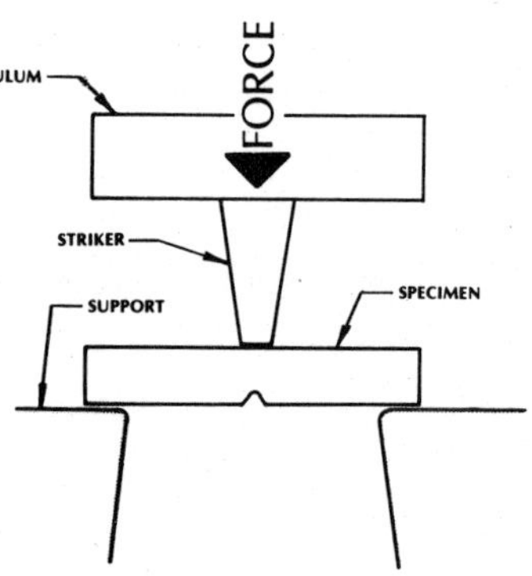

C

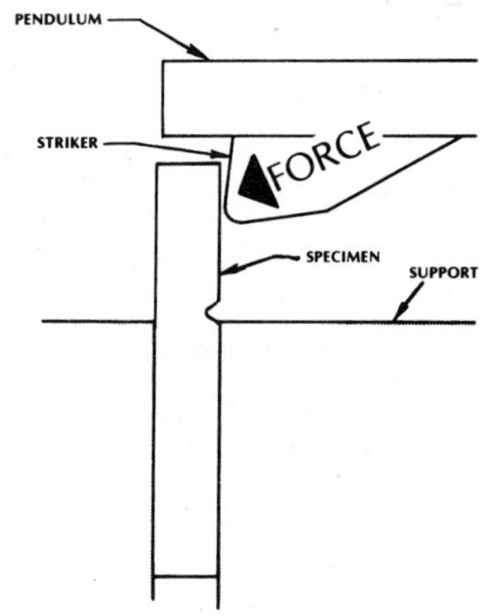

D

REVIEW QUESTIONS 35

Reading Welding Symbols

True-False

T F 1. The American Welding Society (AWS) has standardized weld symbols used in industry today.

T F 2. The two locations that a weld symbol can be placed are the arrow side and the opposite side.

T F 3. The width of a fillet weld is shown to the left of the symbol.

T F 4. The member that is to be beveled on beveled joints is shown with an arrow, with a definite break, pointing toward it.

T F 5. Intermittent fillet welds are shown to the left and below the weld symbol.

T F 6. Size is not shown for single and symmetrical double-groove welds with complete penetration.

T F 7. The size of a spot weld is the diameter of the weld and is located to the left of the symbol.

T F 8. Seam welds are dimensioned by length or width of the weld.

T F 9. The perpendicular leg of the weld symbol always appears on the left side.

T F 10. Combined weld symbols are shown next to each other on the reference line of the arrow.

Multiple Choice

__________ 1. The __________ is included only when a definite welding specification or procedure must be indicated.
- A. surface contour
- B. weld around symbol
- C. back weld
- D. reference tail
- E. none of the above

__________ 2. __________ is not a resistance welding process.
- A. Projection welding (RPW)
- B. Flash welding (FW)
- C. Resistance seam welding (RSEW)
- D. Forge welding (FOW)
- E. none of the above

__________ 3. Plug welds are dimensioned on the weld symbol by __________.
- A. size
- B. depth
- C. angle
- D. pitch
- E. all of the above

__________ 4. __________ is a type of weld.
- A. Plug
- B. Fillet
- C. Groove
- D. all of the above
- E. none of the above

_____ **5.** The size of a _____ weld with unequal legs is given in parentheses to the left of the weld symbol.
A. groove
B. fillet
C. butt
D. flange
E. all of the above

_____ **6.** The number in the weld symbol shown indicates the _____.
A. size of contour
B. depth of penetration
C. root opening size
D. size of bevel face
E. none of the above

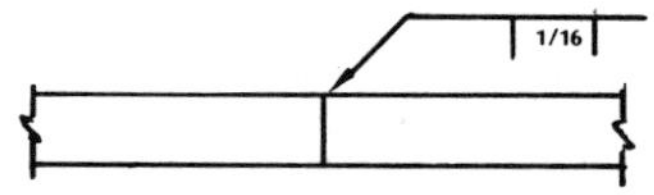

_____ **7.** The number in the weld symbol shown indicates _____.
A. root opening
B. groove face penetration
C. bevel angle
D. effective throat size
E. none of the above

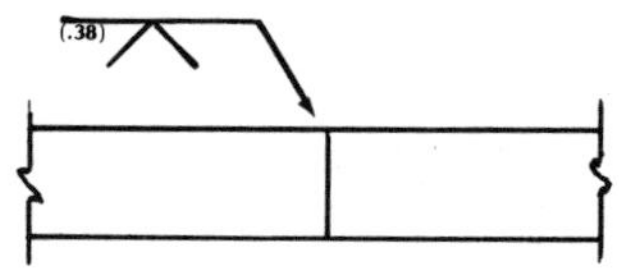

_____ **8.** The _____ joint shown is used for brazing.
A. groove
B. square
C. V
D. scarf
E. none of the above

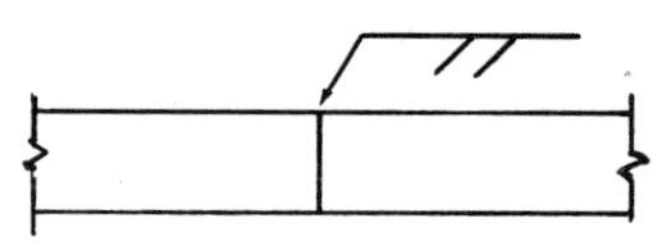

Matching

Identify the basic types of joints shown.

_____ **1.** Corner
_____ **2.** T
_____ **3.** Lap
_____ **4.** Edge
_____ **5.** Butt

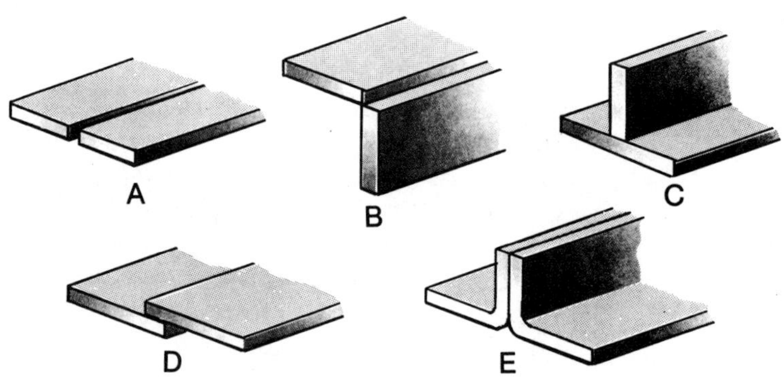

Identify the parts of the base of the weld symbol shown.

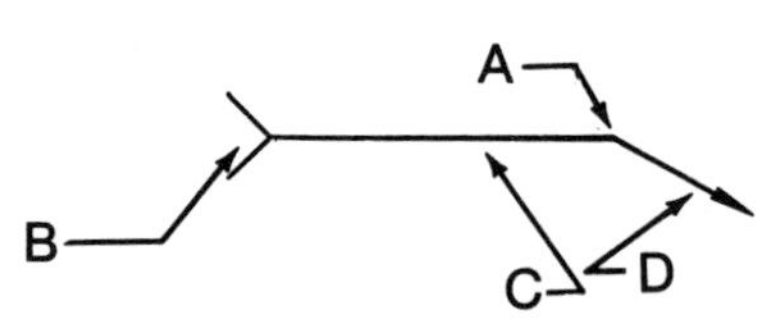

_____ **6.** Type of weld above and below line
_____ **7.** Location of weld
_____ **8.** Reference data
_____ **9.** Location of field weld or weld all around symbol

Identify the weld symbols shown.

__________ **10.** Fillet
__________ **11.** Plug or slot
__________ **12.** Spot or projection
__________ **13.** Seam
__________ **14.** Flange

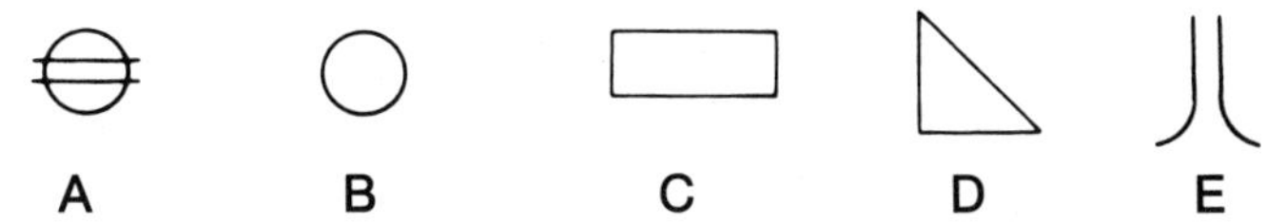

__________ **15.** Back or backing
__________ **16.** Melt-thru
__________ **17.** Surfacing
__________ **18.** Square
__________ **19.** V

__________ **20.** Bevel
__________ **21.** U
__________ **22.** J
__________ **23.** Flare-V
__________ **24.** Flare-bevel

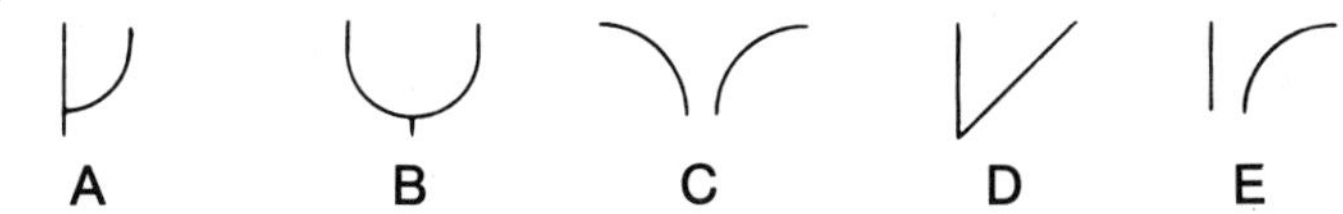

__________ **25.** Weld around
__________ **26.** Field weld
__________ **27.** Flush contour
__________ **28.** Convex contour
__________ **29.** Concave contour

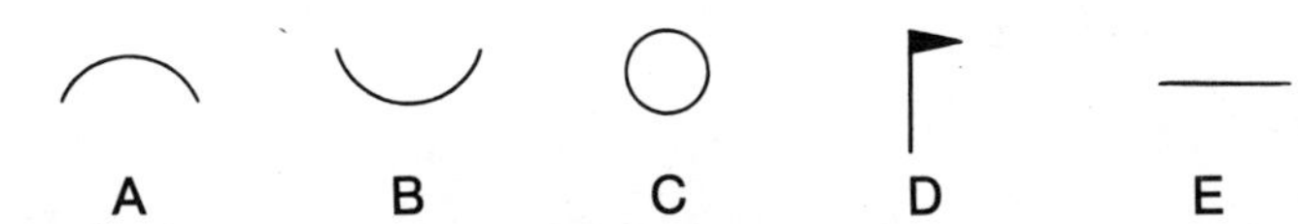

Identify the parts of the master weld symbols shown.

__________ **30.** Finish symbol
__________ **31.** Contour symbol
__________ **32.** Root opening; depth of filling for plug and slot welds
__________ **33.** Size; size or strength for resistance welds
__________ **34.** Specification, process references

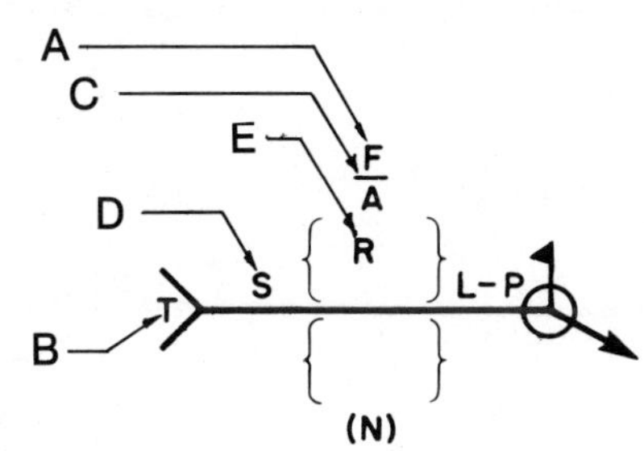

__________ **35.** Tail
__________ **36.** Basic weld symbol
__________ **37.** Groove angle; included angle of countersink for plug welds
__________ **38.** Length of weld
__________ **39.** Pitch (center-to-center spacing) of welds

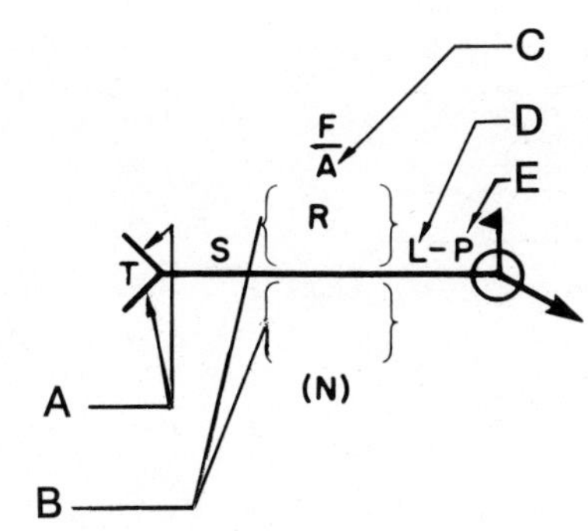

__________ **40.** Field weld symbol
__________ **41.** Weld all around symbol
__________ **42.** Number of spot or projection welds

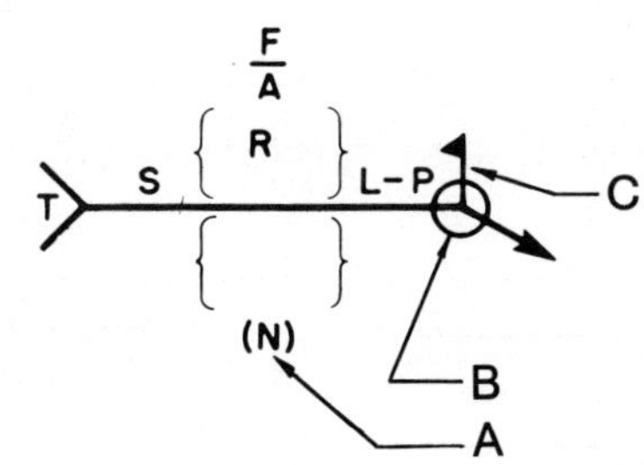

REVIEW QUESTIONS 36

Certification of Welders

True-False

T F 1. Degrees of welding competencies may be expressed in codes, standards, or specifications.

T F 2. Manufacturers generally have their own testing programs to qualify welders for specific tasks.

T F 3. Once a welder is certified for a welding job, additional qualifying tests are not required for a new welding assignment.

T F 4. Qualifying tests are performed on specimens (coupons) of the same material used in the product involved.

T F 5. A welder who successfully completes a certifying test can perform the same task at another manufacturer.

Multiple Choice

________ 1. Certification of a welder involves the testing of weld specimens using one or more or the following tests.

A. Radiographic
B. Fillet weld
C. Tension
D. Guided bend
E. all of the above

________ 2. The ________ electrode is a specified requirement for structural welding when using the GMAW process.

A. E-60XX
B. E-70XX
C. E-60S
D. E-70S
E. none of the above

________ 3. A ________ consists of a set of regulations covering permissible materials, service limitations, fabrication, inspection, testing procedures and qualifications of welding operators.

A. specification
B. certification
C. standard
D. code
E. all of the above

________ 4. ________ are specific regulations that cover the quality of a particular product to be fabricated by welding.

A. Qualification requirements
B. Standards
C. Codes
D. Specifications
E. all of the above

________ 5. ________ are specific descriptions of fabricating procedure.

A. Certification requirements
B. Standards
C. Codes
D. Specifications
E. none of the above

Matching

Identify the weld test positions shown established by the American Welding Society.

__________ **1.** 1G

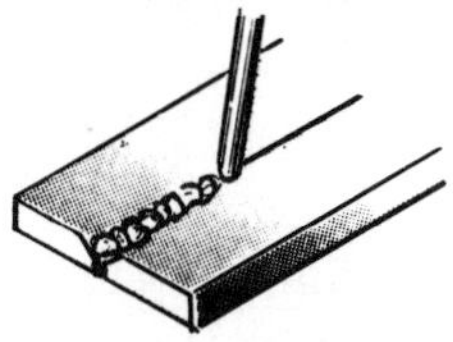

__________ **2.** 2G

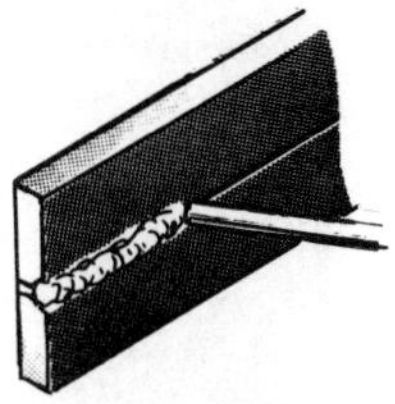

__________ **3.** 3G

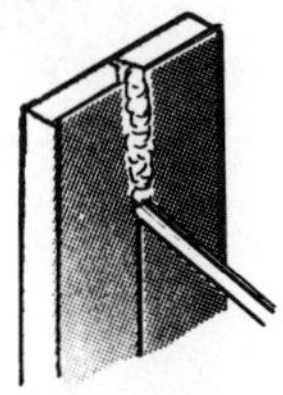

__________ **4.** 4G

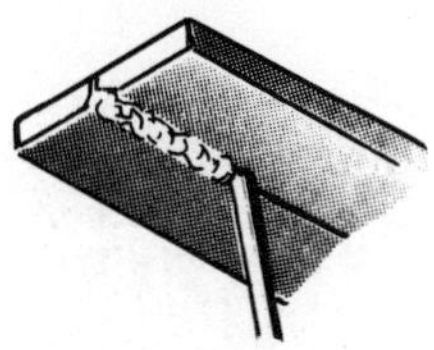

A. specifies vertical position for groove welds
B. specifies horizontal position for groove welds
C. specifies overhead position for groove welds
D. specifies flat position groove welds

Identify the defective fillet weld profiles shown.

__________ **5.** Excessive undercut
__________ **6.** Overlap
__________ **7.** Insufficient throat
__________ **8.** Insufficient leg
__________ **9.** Excessive convexity

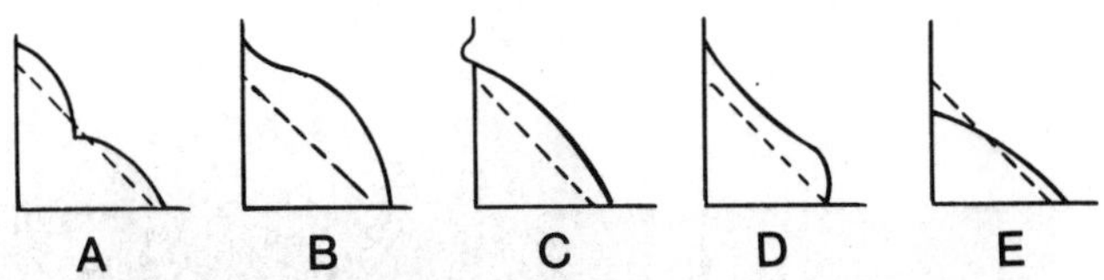

Identify the butt weld profiles shown.

__________ **10.** Insufficient throat
__________ **11.** Excessive undercut
__________ **12.** Acceptable butt weld profile
__________ **13.** Overlap
__________ **14.** Excessive convexity

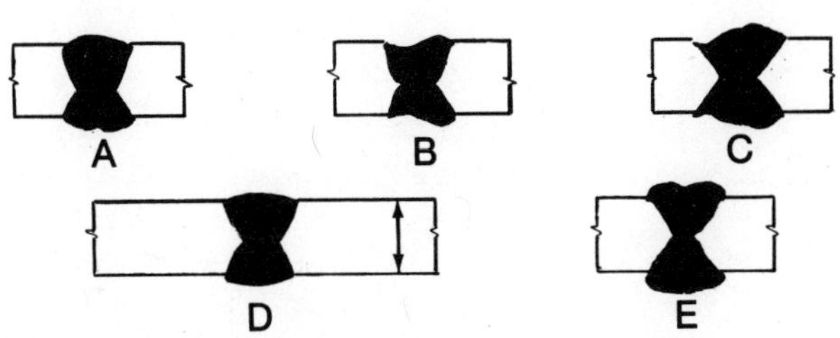

Identify the test specimen specifications for pipe and tubing shown.

__________ **15.** Tensile
__________ **16.** Root bend
__________ **17.** Face bend
__________ **18.** Top of pipe for 5G and 6G positions
__________ **19.** Angle on-center between test specimens

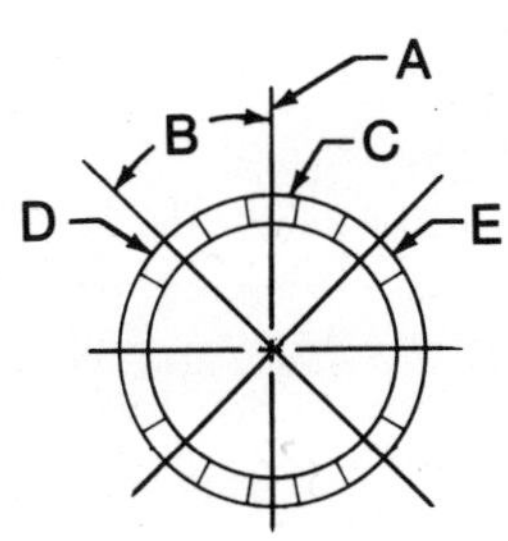

Identify the AWS pipe welding test positions shown.

_______ **20.** 1G
_______ **21.** 2G
_______ **22.** 5G
_______ **23.** 6G

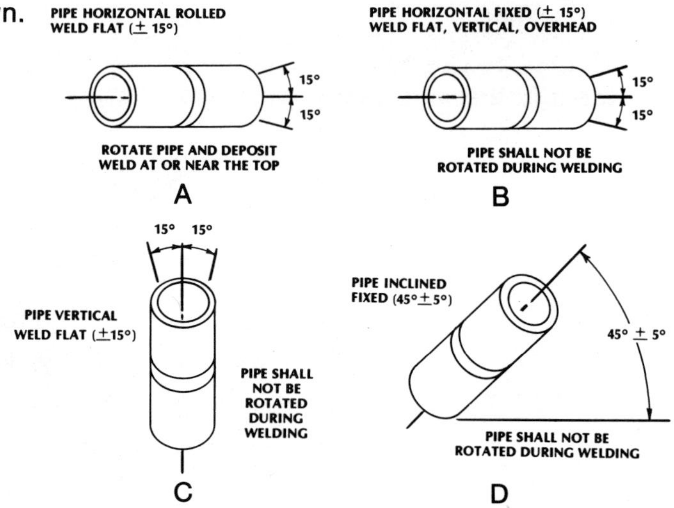

Identify the following qualification levels for pipe and tubing welding as established by the American Welding Society.

_______ **24.** AR-1
_______ **25.** AR-2
_______ **26.** AR-3

A. applies to systems where a nominal degree of weld quality is required
B. applies to systems where the highest level of quality is required
C. applies to systems where a high degree of weld quality is required, such as petroleum, gas, or chemical systems

TECH-CHEK ✓ 1

Name ______________________

Class ____________ Date ____________

Score ____________

An Essential Skill

True-False

T F **1.** Welding is a means of joining materials at a faster rate than other joining methods and is less expensive.

T F **2.** Forge welding was first used in industry in the late 1800s.

T F **3.** Skilled welders are certified for specific welding jobs.

T F **4.** Inert gas welding was developed to weld aluminum and magnesium on World War II airplanes.

T F **5.** The gas metal-arc welding process does not require a shielding gas.

T F **6.** Welding as an occupation does not require specific training.

T F **7.** Welding processes are used in the mass production of automobiles.

T F **8.** Acetylene may be substituted with mapp or natural gas in some welding applications.

T F **9.** The shielded metal-arc welding process deposits molten metal from the tip of the electrode into the weld joint.

T F **10.** The gas tungsten-arc welding process was used before the oxyacetylene welding process was developed.

Multiple Choice

D **1.** The ____________ welding process is used primarily for production welding.

A. forge
B. gas shielded-arc
C. gas metal-arc
D. resistance
E. all of the above

E **2.** The ____________ welding process is known for its mobility and flexibility.

A. forge
B. shielded metal-arc
C. gas metal-arc
D. resistance
E. none of the above

E **3.** The ____________ welding process uses a shielding gas to protect the weld area.

A. gas metal-arc
B. gas shielded-arc
C. gas tungsten arc
D. inert gas
E. all of the above

C **4.** The ____________ welding process uses a flame to generate the heat needed for welding.

A. gas metal-arc
B. gas tungsten-arc
C. oxyacetylene
D. resistance
E. none of the above

B

______ **5.** The two main types of arc welding processes used today are ______ and ______.

A. resistance, oxyacetylene
B. shielded metal-arc, gas shielded arc
C. robotic, resistance
D. all of the above
E. none of the above

______ **6.** The ______ welding process is commonly used in industry today.

A. resistance
B. shielded metal-arc
C. gas metal-arc
D. gas tungsten
E. all of the above

______ **7.** ______ cutting is used to cut various metals to a desired shape.

A. Resistance
B. Inert gas
C. Shielded metal-arc
D. Flame or oxygen
E. none of the above

______ **8.** A course in ______ is helpful to a person considering a career as a welder.

A. mathematics
B. mechanical drawing
C. blueprint reading
D. metals
E. all of the above

______ **9.** Identify the arrow side of the weld joint shown.

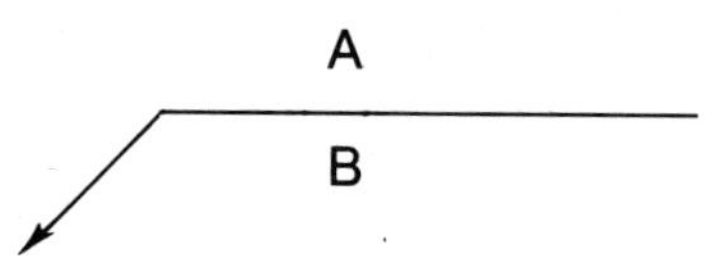

A. A
B. B
C. either
D. both
E. none of the above

______ **10.** Identify the other side of the welding symbol shown.

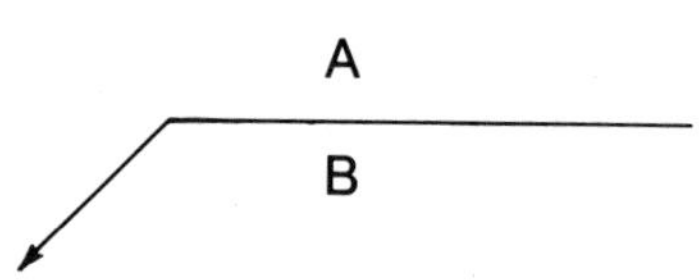

A. A
B. B
C. either
D. both
E. none of the above

TECH-CHEK ✓ 2

Welding Safety

Name ______________________

Class ____________ Date ____________

Score ____________

True-False

T F 1. When welding is completed the welding machine should be turned off.

T F 2. A welding machine that is not properly grounded can cause a severe shock.

T F 3. Operating welding cables above their rated capacity results in faster welding speeds.

T F 4. Oxygen should be used to ventilate a closed container before welding.

T F 5. A fire extinguisher should be accessible in locations where welding is done.

T F 6. When welding or cutting a container, a vent is required to allow for the release of air pressure or steam.

T F 7. Before striking an arc, check to make sure that anyone nearby is adequately protected from the arc flash.

Multiple Choice

__________ 1. Which general safety rules should be followed when welding?
- A. Report all injuries.
- B. Protect the body properly to prevent injury.
- C. Never attempt to operate any equipment without proper instruction.
- D. Always practice good safety habits.
- E. all of the above

__________ 2. A respirator is used when welding metals that give off __________ fumes.
- A. electrode
- B. oxygen
- C. toxic
- D. acetylene
- E. all of the above

__________ 3. When using oxygen and acetylene cylinders open the valves __________.
- A. quickly
- B. slowly
- C. with a pair of pliers
- D. all of the above
- E. none of the above

__________ 4. When repairing a welding machine, the power disconnect must be __________.
- A. turned off
- B. connected
- C. loaded
- D. all of the above
- E. none of the above

__________ 5. All welding equipment must be installed according to provisions of the __________.
- A. operating instructions
- B. owner's manual
- C. National Electrical Code®
- D. supervisor
- E. none of the above

_______ 6. Accidents occur because of _______.
A. indifference to regulations
B. lack of information
C. carelessness
D. all of the above
E. none of the above

_______ 7. _______ should never come in contact with unalloyed copper except in a torch.
A. Carbon dioxide
B. Air
C. Argon
D. Acetylene
E. none of the above

_______ 8. What is the meaning of the welding symbol shown?

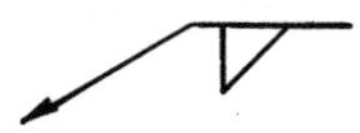

A. Fillet weld arrow side
B. Fillet weld other side
C. Fillet weld both sides
D. Slot weld arrow side
E. none of the above

Matching

_______ 1. Do not weld on hollow castings unless they have been properly _______.
_______ 2. The polarity switch should not be _______ when the welding machine is under load.
_______ 3. Repairs to welding equipment should be made with the _______ off.
_______ 4. Pieces of metal that have been _______ should be allowed to cool before they are picked up.
_______ 5. Before a container is welded or cut, the substance that was held in the container must be positively _______.

A. welded
B. identified
C. vented
D. power
E. switched

_______ 6. The welding area must be properly _______ to prevent injury from welding fumes.
_______ 7. Operating with currents above their _______ capacity causes overheating.
_______ 8. Only _______ personnel should designate the cleaning procedure before welding or cutting containers that have held hazardous substances.
_______ 9. The welding machine must be properly _______ to prevent injury from electrical shock.
_______ 10. Assume that all _______ objects are hot before touching them.

A. welded
B. ventilated
C. rated
D. grounded
E. qualified

TECH-CHEK ✓ 3

Name ____________________

Class __________ Date __________

Score __________

Welding Metallurgy

True-False

T F 1. Metal pieces being welded are held in a fixed position with a jig.

T F 2. The back-step welding technique is used to increase the heat applied to the weld area.

T F 3. Cryogenic properties of metal refer to the characteristics of the metal when subjected to a low temperature environment.

T F 4. The coefficient of expansion of a given metal is the amount of expansion when the metal is heated.

T F 5. Toughness of a given metal refers to strength combined with ductility.

T F 6. The brittleness of a given metal is determined by how far it will stretch before breaking.

T F 7. Hardness of a metal is the amount of resistance to indentation or penetration.

T F 8. The ductility of a given metal is its ability to stretch, bend, or twist without breaking or cracking.

T F 9. The property of metal that resists forces acting to pull the metal apart is its tensile strength.

T F 10. Compressive strength is the ability of a metal to withstand forces to cause a member to twist.

T F 11. Stress is the internal resistance the metal offers.

T F 12. The modulus of elasticity is the amount of contraction that occurs when heating the metal.

T F 13. The ability of a metal to return to its original shape after distortion is its elasticity.

T F 14. The elastic limit is the last point at which a metal may be stretched and still return to its original shape in its original condition.

T F 15. Strain is the deformation of a metal that results from stress.

T F 16. Distortion is caused from expansion and contraction rates in the metal being welded.

T F 17. For most butt welds, allow 1/2″ for each foot in length of the weld.

Matching

________ 1. ____________ temperature is the temperature of transformation for a given metal.

________ 2. ____________ is a process of softening metal to be machined.

________ 3. ____________ properties of metals involve corrosion, oxidation, and reduction.

________ 4. ____________ is a process of hardening steel by the absorption of carbon on its surface.

________ 5. ____________ properties refer to the behavior of metals when subjected to applied loads.

A. Annealing
B. Chemical
C. Case hardening
D. Mechanical
E. Critical

_______ 6. The grain structure of metal _______ when heated.

_______ 7. A welded joint _______ as it cools.

_______ 8. Correct preheat and postweld heat treatment reduce _______ built up from welding.

_______ 9. Physical _______ of metals include melting point, thermal conductivity, and grain structure.

_______ 10. The surface of metal _______ using the cyaniding process.

A. contracts
B. changes
C. properties
D. hardens
E. stresses

Multiple Choice

_______ 1. Metal _______ when heated and _______ when cooled.
   A. expands, contracts
   B. contracts, expands
   C. shrinks, spreads
   D. stretches, become brittle
   E. none of the above

_______ 2. Which of the following is not used to control distortion?
   A. Back-step welding
   B. Intermittent welding
   C. Peening the weld
   D. Inclusion welding
   E. all of the above

_______ 3. _______ is not a mechanical property of a metal.
   A. Tensile strength
   B. Shear strength
   C. Compressive strength
   D. Torsional strength
   E. none of the above

_______ 4. Which of the following is not a welding defect?
   A. Segregation
   B. Blowholes
   C. Porosity
   D. Cryogenic properties
   E. none of the above

_______ 5. What is the meaning of the welding symbol shown?

   A. Fillet weld arrow side
   B. Fillet weld other side
   C. Fillet weld both sides
   D. Slot weld arrow side
   E. none of the above

TECH-CHEK ✓ 4

Name ____________________

Class __________ Date __________

Score __________

Joint Design and Welding Terms

True-False

T F **1.** The distance from toe to toe across the face of the weld is the face.

T F **2.** The square T-joint does not require edge preparation.

T F **3.** A weld pass refers to depth of fusion left by the shape of the deposited bead.

T F **4.** The base metal is the metal on which a weld is deposited.

T F **5.** A pipe weld commonly requires a 75° groove angle.

Multiple Choice

_______ **1.** ____________ is a consideration taken to determine the weld joint best suited for a given job.

A. Thickness of the plates
B. Direction of the load
C. Whether a load is applied steadily, suddenly, or variably
D. Whether a load is in tension or in compression and if bending, fatigue, or impact stresses will be encountered.
E. all of the above

_______ **2.** ____________ is not a welding position.

A. Full-open
B. Flat
C. Horizontal
D. Vertical
E. Overhead

Matching

Questions 1–3

Identify the parts of the weld shown:

_______ **1.** base metal
_______ **2.** penetration
_______ **3.** bead

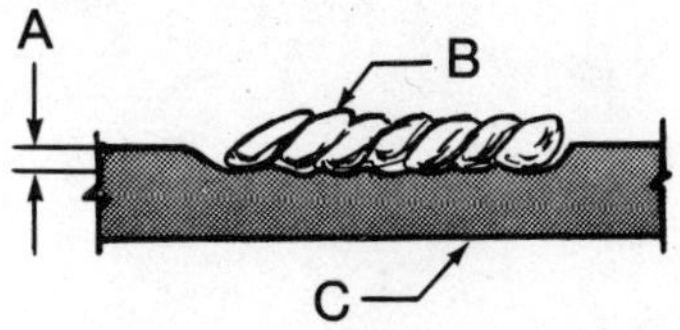

Questions 4–8

Identify the parts of the T-joint shown:

_______ **4.** leg
_______ **5.** root
_______ **6.** throat
_______ **7.** face
_______ **8.** toe

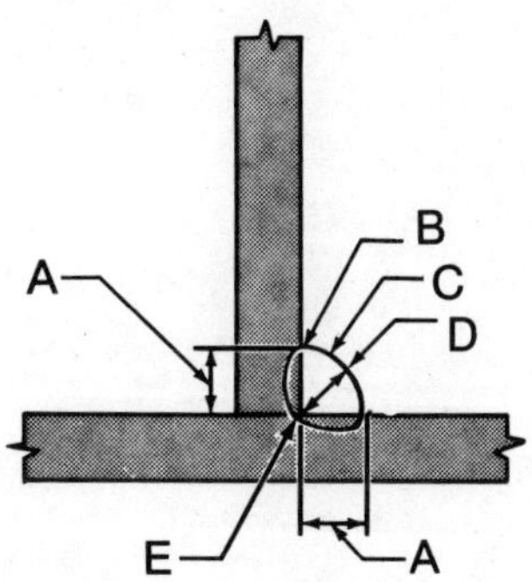

Questions 9–12

Identify the parts of the groove weld shown:

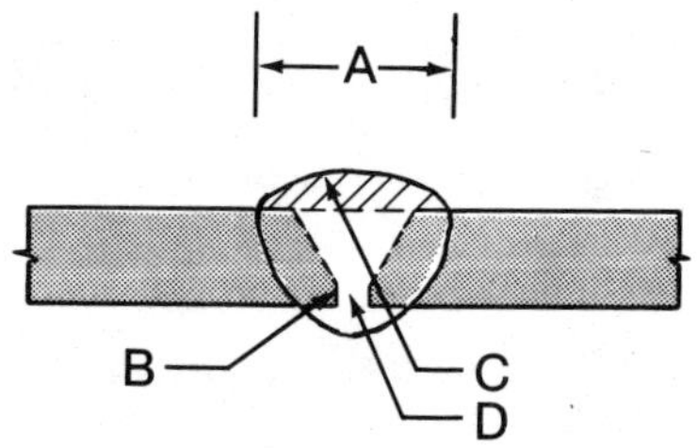

__________ **9.** root face
__________ **10.** reinforcement
__________ **11.** root opening
__________ **12.** weld width

Questions 13–16

Identify the welding positions shown:

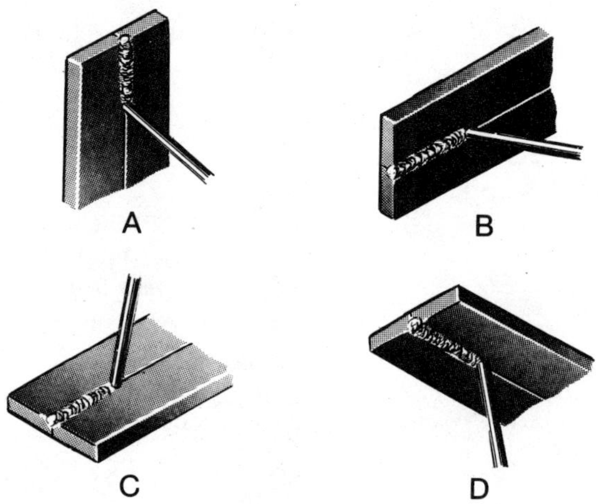

__________ **13.** flat
__________ **14.** horizontal
__________ **15.** vertical
__________ **16.** overhead

TECH-CHEK ✓ 5

Name ____________________

Class __________ Date __________

Score __________

Equipment

True-False

T F **1.** Left-hand threads are on all connectors for acetylene gas.

T F **2.** Oxygen cylinders are packed with a porous material to store oxygen safely.

T F **3.** Cylinders should be chained in a vertical position.

T F **4.** Regulators must be carefully handled to prevent possible damage.

T F **5.** Shade numbers commonly used in oxyacetylene welding goggles are 10, 11, and 12.

T F **6.** Regulators should be repaired by qualified personnel only.

T F **7.** Right-hand threads are always used on connections for acetylene hoses.

T F **8.** Acetylene becomes unstable if compressed to more than 15 psi.

T F **9.** A match should be used to ignite the oxyacetylene flame.

T F **10.** Acetone absorbs large amounts of acetylene, allowing storage of gas at a safe pressure.

T F **11.** The oxygen regulator gauge is marked with a warning color above the 15 psi point.

T F **12.** Regulators should be greased or oiled occasionally to guarantee smooth operation.

T F **13.** The higher the welding torch tip number is, the larger the tip opening is.

Matching

________ **1.** A protector cap guards against damage to the ____________.

________ **2.** The ____________ is used to increase or decrease the gas pressure delivered to the torch from the regulator.

________ **3.** Two or more cylinders are connected in line in an oxyacetylene ________.

________ **4.** A __________ controls the flow of gas from the cylinder.

________ **5.** A __________ controls the flow of gas at the welding torch.

A. manifold system
B. regulator
C. adjusting screw
D. needle valve
E. cylinder valve

________ **6.** A ________ safely ignites the torch flame.

________ **7.** An oxygen cylinder contains ________ cubic feet of oxygen.

________ **8.** An oxygen cylinder is pressurized to _____ pounds.

________ **9.** A ______ mixes oxygen and acetylene.

________ **10.** A ________ prevents a flashback from reaching the manifold system.

A. welding torch
B. striker
C. 244
D. flash arrestor
E. 2,200

Identify the parts of the oxygen cylinder shown.

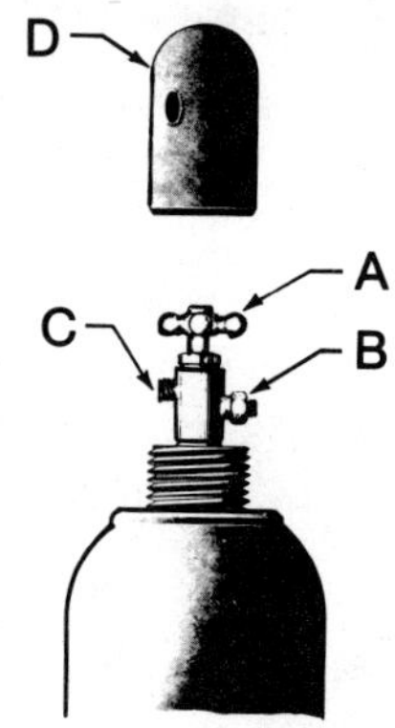

__________ **11.** Outlet nozzle

__________ **12.** Valve handwheel

__________ **13.** Protector cap

__________ **14.** Safety nut

TECH-CHEK ✓ 6

Name ____________________

Class ____________ Date ____________

Score ____________

Setting-Up and Operating

True-False

T F **1.** No pressure should be left in the welding hoses when the welding unit is shut down.

T F **2.** A neutral flame results from a one-to-one mixture of oxygen and acetylene.

T F **3.** An oxidizing flame has a slight feather extending from the inner cone of the flame.

T F **4.** The correct torch tip number to use is determined by the thickness of the metal being welded.

T F **5.** Welding hoses are purged by opening the cylinder valves.

T F **6.** A torch should never be relighted by touching it against the hot metal of the weld.

T F **7.** Gas waste is avoided by opening a cylinder valve quickly.

T F **8.** Check valves are installed to prevent the reverse flow of gases.

T F **9.** The connecting nuts on the acetylene hose have left-handed threads.

T F **10.** Popping is caused by insufficient gas flow to the welding torch.

Multiple Choice

_______ **1.** Which of the following is not a recommended procedure when using oxygen and acetylene cylinders?
- A. Chain down the cylinders.
- B. Crack the valves to remove any foreign matter.
- C. Open the acetylene cylinder one-half turn.
- D. Open the oxygen cylinder one-half turn.
- E. none of the above

_______ **2.** Which of the following is not a recommended procedure for lighting a welding torch?
- A. Use a striker to light a torch.
- B. Never reach over another person for a light.
- C. Always use a striker to relight a flame.
- D. Hold the tip upward.
- E. none of the above

Matching

Identify the following flames shown:

_______ **1.** Oxidizing flame

_______ **2.** Carburizing flame

_______ **3.** Neutral flame

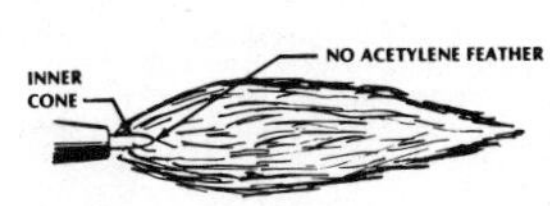

A

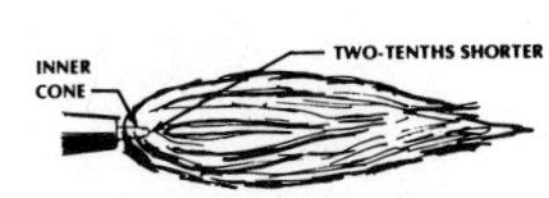

B

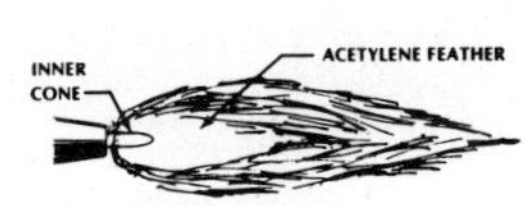

C

_______ **4.** All new welding apparatus must be _______ for leaks.

_______ **5.** Check valves are _______ between the welding torch and hose.

_______ **6.** Cylinders should be _______ to a stationary object to prevent accidental tipping.

_______ **7.** To remove any particles of dirt, the cylinder valves should be _______ before mounting the regulators.

_______ **8.** The welding hoses are _______ of residual gases before attaching the welding torch.

A. chained
B. purged
C. cracked
D. tested
E. mounted

_______ **9.** The _______ needle valve is opened first when lighting the torch.

_______ **10.** Testing for leaks requires the use of _______.

_______ **11.** When lighting the torch, the _______ should point downward.

_______ **12.** Stand to one side of the regulator when opening a _______.

_______ **13.** Close the _______ needle valve first when shutting off the welding torch.

A. cylinder valve
B. welding tip
C. acetylene
D. oxygen
E. soapy water

Identify the flames shown.

_______ **14.** Excess of acetylene
_______ **15.** Neutral flame
_______ **16.** Acetylene burning in atmosphere

A

B

C

TECH-CHEK ✓ 7

Name ______________________

Class ____________ Date ____________

Flat Position

Score ____________

True-False

T F **1.** The welding torch should be positioned at 45° to the metal when running beads.

T F **2.** The filler rod should be dipped into the middle of the molten puddle to prevent the rod from sticking to the base metal.

T F **3.** A hole in the weld joint results from holding the flame in one location too long.

T F **4.** Tack welds made before depositing the bead prevent expansion and contraction forces from distorting the weld joint.

T F **5.** The welding torch can be held like a hammer or like a pencil depending on the preference of the welder.

T F **6.** Different filler rods are required for welding on alloyed steels.

T F **7.** The travel speed of a torch must be reduced if the puddle becomes too large.

T F **8.** When progressive spacing is used, the root opening is smaller at the beginning of the weld than at the end of the weld.

T F **9.** With correct penetration on a butt joint, fusion of the weld metal will show to the bottom of the joint.

T F **10.** A corner joint can be welded without adding filler rod.

Multiple Choice

________ **1.** When welding a T-joint in the flat position, the torch is angled at a ____________° to the work.

A. 90
B. 75
C. 60
D. 45
E. 30

________ **2.** When welding a lap joint in flat position, ____________ to avoid overheating.

A. lift the torch 3″ above the work
B. direct more heat to the top plate
C. direct more heat to the bottom plate
D. move the torch more slowly
E. all of the above

________ **3.** The correct diameter of filler rod is approximately ____________ the thickness of the base metal.

A. one-half
B. twice
C. equal to
D. three times
E. none of the above

________ **4.** Carrying a puddle without filler rod requires that the inner cone of the flame be held approximately ____________ from the work.

A. 1/16″
B. 1/8″
C. 3/16″
D. 1/4″
E. none of the above

_____ **5.** When welding, heat from the torch should be concentrated on the _____.

A. filler rod
B. base metal
C. area around the weld joint
D. end of the weld joint
E. none of the above

_____ **6.** What is the meaning of the welding symbol shown?

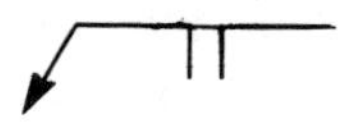

A. Fillet weld arrow side
B. Slot weld other side
C. Butt weld arrow side
D. Butt weld other side
E. none of the above

_____ **7.** What is the meaning of the welding symbol shown?

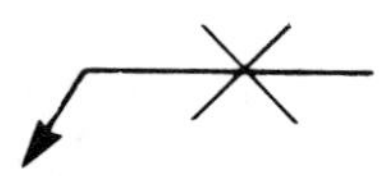

A. Double V-groove weld
B. V-groove weld arrow side
C. Double fillet
D. Double butt
E. none of the above

Matching

_____ **1.** If the _____ is too slow, the torch may burn through the metal.

_____ **2.** Heat from the torch should be directed to the _____.

_____ **3.** The _____ should be dipped into the middle of the weld puddle.

_____ **4.** A circular or semicircular _____ is used to maintain an adequate puddle.

_____ **5.** The _____ is advanced about 1/16" with each complete motion of the torch.

A. filler rod
B. travel speed
C. torch motion
D. puddle
E. base metal

TECH-CHEK ✓ 8

Name ______________________

Class ____________ Date ____________

Score ____________

Other Welding Positions

True-False

T F 1. Gravitational pull of molten metal is the main obstacle preventing a sound weld when welding in positions other than flat.

T F 2. Molten metal has cohesive qualities as long as the puddle does not get too large.

T F 3. Tack welding the metal pieces for a horizontal butt joint requires that the edges be tightly butted against each other for maximum penetration.

T F 4. Filler rod added closer to the vertical plate prevents undercutting when welding a horizontal T-joint.

T F 5. Weld pieces can be mounted in a jig when practicing welds in the horizontal, vertical, and overhead positions.

Multiple Choice

__________ 1. Heat directed to the weld joint should be ____________ to prevent the weld puddle from becoming too large when welding in the overhead position.

A. increased
B. reduced
C. kept the same
D. raised slightly
E. none of the above

__________ 2. ____________ controls the shape of the deposited bead when welding in positions other than flat.

A. Proper torch manipulation
B. Keeping the flame at the proper distance
C. Adding filler rod to the puddle where it will prevent undercutting
D. all of the above
E. none of the above

__________ 3. On overhead welds, the filler rod should be moved slowly in a ____________ or swinging motion.

A. up and down
B. straight
C. triangular
D. circular
E. none of the above

__________ 4. When horizontal welding a butt joint, the flame should be directed more on the ____________.

A. weld puddle
B. bottom plate
C. top plate
D. end of the weld crater
E. all of the above

__________ 5. The overhead welding position ____________.

A. is the most difficult of the main welding positions
B. requires the correct flame and proper torch manipulation
C. is possible as long as the weld puddle is not too large
D. is practiced by placing the weld pieces in a suitable jig
E. all of the above

_______ **6.** What is the meaning of the welding symbol shown?

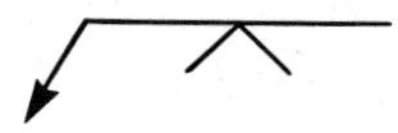

A. Groove weld arrow side
B. Single V-groove weld arrow side
C. Fillet weld arrow side
D. Butt weld other side
E. Fillet weld other side

_______ **7.** What is the meaning of the welding symbol shown?

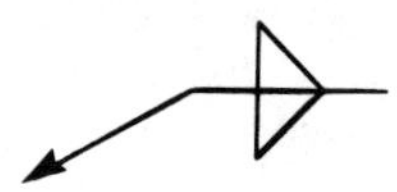

A. Fillet weld arrow side
B. Fillet weld other side
C. Fillet weld both sides
D. Slot weld arrow side
E. none of the above

_______ **8.** What is the meaning of the welding symbol shown?

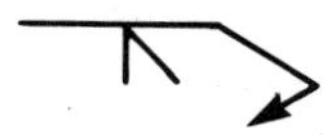

A. Bevel-groove arrow side, 90° face left side
B. Groove other side
C. Bevel-groove other side, 90° face left side
D. Groove arrow side
E. none of the above

TECH-CHEK ✓ 9

Heavy Steel Plate

Name ______________________

Class ____________ Date ____________

Score ____________

True-False

T F 1. In the backhand welding technique, the welding torch precedes the filler rod.

T F 2. Metal over $^1/_8$″ thick is designated as plate.

T F 3. A heat treating process must be used to restore the hardness of high-carbon steel after it has beeen welded.

T F 4. The forehand welding technique is recommended for welding heavy plate.

T F 5. A flame with a slight excess of acetylene is used for welding high-carbon steel.

Multiple Choice

_________ 1. Filler rods that have a __________ composition compared to the base metal should be used when welding heavy sections of high-carbon steel.
- A. higher nickel
- B. low-carbon
- C. tungsten carbide
- D. similar
- E. none of the above

_________ 2. When welding a butt joint, metals with a thickness exceeding __________″ must be beveled.
- A. $^1/_{16}$
- B. $^1/_8$
- C. $^3/_{16}$
- D. $^1/_4$
- E. none of the above

_________ 3. The recommended included angle for a single V-butt joint is __________°.
- A. 20
- B. 45
- C. 120
- D. 60
- E. none of the above

_________ 4. Which of the following is true when welding plates ½″ or more in thickness?
- A. Prepare the edges as a single V.
- B. Prepare the edges as a double V.
- C. Use a single pass to obtain maximum penetration.
- D. Use a forehand welding technique.
- E. none of the above

_________ 5. Which of the following is true when welding heavy steel plate using the oxyacetylene process?
- A. The oxyacetylene process is faster than other processes.
- B. Heavy steel plate is frequently welded using the oxyacetylene process.
- C. The torch angles used for welding heavy steel plates are the same as those used for light gauge materials.
- D. The joints must be carefully prepared to allow for sufficient penetration.
- E. all of the above

_______ **6.** What is the meaning of the welding symbol shown?

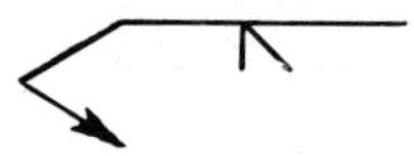

A. Bevel-groove weld arrow side
B. Bevel-groove weld other side
C. Groove weld
D. Fillet weld other side
E. none of the above

Matching

_______ **1.** The correct _______ is applied before welding high-carbon steel plate.
_______ **2.** A _______ is obtained on the torch before welding high-carbon steel plate.
_______ **3.** A _______ is used to test the weld.
_______ **4.** A _______ is used to fill the V of the weld when welding high-carbon steel plate.
_______ **5.** The _______ is used when welding heavy plate.

A. guided-bend tester
B. preheat technique
C. high-carbon filler rod
D. backhand welding technique
E. carburizing flame

_______ **6.** To ensure maximum penetration, _______ is required when welding heavy steel plate.
_______ **7.** Before adding filler rod make certain that the sides are in a molten state down to _______.
_______ **8.** The grind marks on the test specimen should run parallel with _______.
_______ **9.** Use _______ with the torch when welding a single V-butt joint.
_______ **10.** Grind off surplus weld metal so that _______ is flush with the top of the plate.

A. the length of the piece
B. the bottom of the V
C. a root opening
D. the top of the weld
E. a semicircular motion

TECH-CHEK ✓ 10

Name ______________________________

Class ______________ Date ______________

Score ______________

Gray Cast Iron

True-False

T F 1. In welding cast iron, flux is essential to keep the molten puddle fluid.

T F 2. Cast iron must be preheated to a bright red before welding.

T F 3. When welding, the torch must be manipulated in a circular motion to distribute the heat evenly.

T F 4. The filler rod used must have the same properties as the base metal being welded.

T F 5. When welding is complete, the casting must be quenched to ensure a strong weld.

Multiple Choice

__________ 1. Which of the following applies when preparing the casting for welding?
- A. The weld area should be cleaned at least 1″ on both sides of the V.
- B. The edges of the casting should be beveled to a 90° included angle.
- C. Carbon back-up blocks are placed underneath the joint to prevent molten cast iron from running through the seam.
- D. The casting should be uniformly preheated.
- E. all of the above

__________ 2. After welding, if the casting is cooled too rapidly, the weld area is likely to turn into ______________.
- A. slag
- B. flux
- C. gray cast iron
- D. white cast iron
- E. none of the above

__________ 3. When adding filler rod to the molten puddle, which of the following is the correct procedure?
- A. Dip the rod in and out of the puddle.
- B. Insert the fluxed end of the filler rod into the puddle.
- C. Position the filler rod on the base metal ahead of the weld puddle.
- D. Apply the flux to the base metal before welding.
- E. all of the above

__________ 4. Preheating of cast iron ____________________.
- A. requires a uniform temperature
- B. minimizes the possibility of cracks developing
- C. can be done with a torch
- D. will equalize expansion and contraction forces
- E. all of the above

__________ 5. Which of the following is not true about flux when used for welding cast iron?
- A. It prevents iron oxides from forming.
- B. It maintains a fluid molten puddle.
- C. The hot filler rod is dipped into the flux.
- D. Flux is not necessary when welding gray cast iron.
- E. all of the above

______ **6.** What is the meaning of the welding symbol shown?

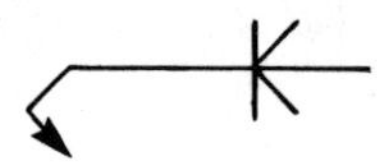

A. Fillet weld both sides
B. Butt joint both sides
C. Groove both sides
D. Bevel arrow side
E. none of the above

Matching

______ **1.** Placing ______ under the joint will prevent molten cast iron from running out of the seam.

______ **2.** Move the ______ in a circular motion to distribute the heat.

______ **3.** Flux is used to prevent ______ from mixing with iron oxide which forms in the weld puddle.

______ **4.** A ______ is used to keep the molten puddle fluid when welding cast iron.

______ **5.** To ensure a strong weld, use the correct ______ to fill the V.

A. carbon back-up blocks
B. flux
C. torch
D. cast-iron filler rod
E. slag

TECH-CHEK ✓ 11

Name ______________________

Class ____________ Date ____________

Score ____________

Aluminum

True-False

T	F	**1.**	Before welding aluminum, preheating is necessary to decrease the effects of expansion and contraction.
T	F	**2.**	The weld area must be cleaned thoroughly before it is welded.
T	F	**3.**	The backhand technique is used when welding aluminum plate.
T	F	**4.**	The recommended preheat temperature for most aluminum welding is 300–500°F.
T	F	**5.**	On thin aluminum, a flange joint is welded without adding filler metal.
T	F	**6.**	Aluminum should be preheated until it is dull red.
T	F	**7.**	Aluminum becomes very soft and weak when heated, requiring special considerations when it is welded.
T	F	**8.**	A neutral or slightly reducing flame is used for all aluminum welding.
T	F	**9.**	The thermal conductivity of aluminum is approximately four times that of steel.
T	F	**10.**	Aluminum has a higher melting point than steel and requires a smaller welding tip.

Multiple Choice

__________ **1.** ______________ when welding 1/16″ to 3/16″ thick aluminum.

A. Lap joints are recommended for a strong joint
B. Flux is not recommended
C. The edges are notched
D. Preheating is not required
E. all of the above

__________ **2.** The welding torch is angled at less than __________° above horizontal when welding a butt joint.

A. 20
B. 30
C. 45
D. 60
E. none of the above

__________ **3.** ______________ filler rods are not used for welding aluminum.

A. 1100
B. 4043
C. 5356
D. all of the above
E. none of the above

__________ **4.** The melting point of pure aluminum is __________°F.

A. 500
B. 800
C. 1220
D. 3000
E. none of the above

_______ **5.** An included angle of _______° is recommended for heavy aluminum plate 3/8" thick or thicker.
A. 10–15
B. 30–45
C. 60–75
D. 75–90
E. 100–120

_______ **6.** A _______ flame is used when welding aluminum.
A. carburizing
B. neutral or slightly reducing
C. oxidizing
D. all of the above
E. none of the above

_______ **7.** _______ is used to prevent oxides from forming in the weld joint.
A. Filler rod
B. A scraping action
C. A carburizing flame
D. Flux
E. none of the above

_______ **8.** Aluminum welds should be made _______ if possible.
A. in three passes
B. in two passes
C. in a single pass
D. using a backhand welding technique
E. none of the above

_______ **9.** What is the meaning of the welding symbol shown?

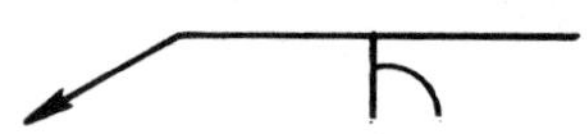

A. J-groove weld other side
B. J-groove weld arrow side
C. J-groove weld both sides
D. V-groove weld other side
E. V-groove weld arrow side

TECH-CHEK ✓ 12

Name ______________________

Class ____________ Date ____________

Score ____________

Machines and Accessories

True-False

T F 1. Alternating current requires that the electrode be positive when welding in reverse polarity.

T F 2. A constant-current welding machine is recommended for shielded metal-arc welding.

T F 3. Amperage is the amount or rate of current flow in a circuit.

T F 4. Resistance is the force that causes current to move in a circuit.

T F 5. Direct current is electrical current that flows in one direction.

T F 6. Safety glasses should be worn under the welding helmet to protect the eyes from slag particles.

T F 7. The actual voltage needed to produce welding current is 18–36 volts.

T F 8. With straight polarity, more heat is directed to the electrode than to the work.

T F 9. Gloves are required to protect hands from hot metal spatter and ultraviolet rays.

T F 10. All exposed metal surfaces of a safe electrode holder are insulated to prevent accidental flashing.

T F 11. The ground connection is connected to the work or bench.

T F 12. In the United States, alternating current is rated at 25 cycles per second.

T F 13. Static electricity refers to electricity at rest or electricity that is not moving.

T F 14. The polarity switch on the welding machine controls the amount of current flowing to the work.

T F 15. Plastic cover plates are used to protect colored lenses in welding helmets.

Multiple Choice

__________ 1. A(n) __________ is any material that allows free passage of electrical current.

A. ampere
B. volt
C. resistance
D. conductor
E. none of the above

__________ 2. Shielded metal-arc welding is sometimes called __________ welding.

A. stick
B. arc
C. metallic arc
D. all of the above
E. none of the above

__________ 3. After the arc is struck, the voltage drops to __________, which is between 18 and 36 volts.

A. dynamic electricity
B. reverse polarity
C. open voltage
D. variable polarity
E. working voltage

_______ **4.** When welding with DC reverse polarity, electrons flow from the _______ to the _______.
- A. work, ground
- B. electrode, work
- C. work, electrode
- D. base metal, ground
- E. none of the above

_______ **5.** A(n) _______ is used to remove slag after the weld is deposited.
- A. cutting torch
- B. chipping hammer
- C. electrode holder
- D. pair of pliers
- E. none of the above

_______ **6.** What is the meaning of the welding symbol shown?

- A. V-groove both sides
- B. Bevel-groove arrow side
- C. V-groove other side
- D. U-groove other side
- E. none of the above

Matching

_______ **1.** The _______ type of welding machine has a sloping volt-amp characteristic.

_______ **2.** The _______ type of welding machine is the least expensive, lightest, and smallest of all welding machines.

_______ **3.** The _______ group of welding machines are used primarily for gas metal-arc welding.

_______ **4.** The _______ type of welding machine can change alternating current into direct current.

_______ **5.** The _______ type of welding machine produces only direct current.

- A. rectifier
- B. transformer
- C. generator
- D. constant-current
- E. constant-potential

_______ **6.** A 60% _______ will put out the rated amperage at the rated voltage for six minutes out of ten.

_______ **7.** A(n) _______ indicates the amount of current flowing in a circuit.

_______ **8.** The opposition of the material in the conductor to the passage of electrical current is referred to as _______.

_______ **9.** When welding with DC current, _______ causes the arc to wander.

_______ **10.** A(n) _______ measures the force of electricity in a circuit.

- A. ammeter
- B. voltmeter
- C. duty cycle
- D. arc blow
- E. resistance

TECH-CHEK ✓ 13

Name ____________________

Class ____________ Date ____________

Score ____________

Selecting Electrodes

True-False

T F **1.** The coating on an electrode acts as a cleansing and deoxidizing agent in the molten crater.

T F **2.** The first two digits of the AWS numerical electrode classification designate the special manufacturer's characteristics.

T F **3.** An E-6010 electrode is designed for use with direct current reverse polarity.

T F **4.** Fast-fill electrodes are designed to be used in horizontal, vertical, and overhead positions.

T F **5.** An E-6013 electrode may be used with AC, DC straight, and DC reverse.

T F **6.** When selecting an electrode, joint design, and fit-up must be considered.

T F **7.** The third digit of an AWS numerical electrode classification indicates the tensile strength of an electrode.

T F **8.** Fast-freeze electrodes can be used with DC straight polarity.

T F **9.** Drying ovens are used to store electrodes subject to damage from moisture in the air.

T F **10.** The prefix E identifies the welding electrode designed for electric arc welding as assigned by the AWS numerical electrode classification.

Multiple Choice

__________ **1.** When DC-EN is recommended for an electrode, __________ current should be used.

A. DC straight
B. DC reverse
C. alternating
D. AC straight
E. none of the above

__________ **2.** An E-6011 electrode should be used with __________ current.

A. DC reverse
B. DC electrode positive
C. alternating
D. all of the above
E. none of the above

__________ **3.** Weld metal deposited should have approximately the same mechanical properties as the __________.

A. flux coating
B. slag
C. base metal
D. shielding gas
E. none of the above

__________ **4.** A __________ electrode is preferred for welding in the overhead position.

A. fast-freeze
B. fast-fill
C. fill-freeze
D. iron powder
E. none of the above

_______ **5.** An _______ is a fast-freeze electrode.

A. E-6010
B. E-6012
C. E-6024
D. E-7024
E. all of the above

_______ **6.** An _______ is a fast-fill electrode.

A. E-6010
B. E-6011
C. E-7024
D. E-6013
E. none of the above

_______ **7.** An _______ is a fill-freeze electrode.

A. E-6010
B. E-6011
C. E-6013
D. E-6024
E. none of the above

_______ **8.** The first two digits in an AWS numerical electrode classification designate the minimum allowable _______ of the weld metal in thousands of pounds per square inch.

A. shielding gas
B. bending strength
C. tensile strength
D. compression strength
E. none of the above

_______ **9.** An electrode designed for welding in any position has the number _______ as the third digit of the AWS numerical electrode classification.

A. 1
B. 2
C. 3
D. 4
E. none of the above

_______ **10.** An electrode designed for welding in flat position or horizontal fillets only, has a number _______ as the third digit of the AWS numerical electrode classification.

A. 1
B. 2
C. 3
D. 4
E. none of the above

_______ **11.** What is the meaning of the welding symbol shown?

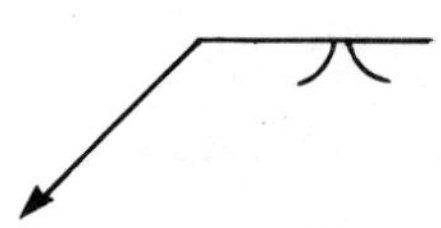

A. Flare bevel groove other side
B. Flare V-groove arrow side
C. Single V-groove
D. Bevel-groove
E. none of the above

TECH-CHEK ✓ 14

Name ______________________

Class ____________ Date ____________

Score ____________

Striking Arcs

True-False

T F **1.** Soapstone is used to mark specific locations on the metal for cutting or welding.

T F **2.** The bare end of the electrode is gripped in the electrode holder to ensure good electrical contact.

T F **3.** A scratching or tapping motion can be used to strike an arc.

T F **4.** In order to strike an arc, the base metal must be free of dirt, rust, and grease.

T F **5.** Welding with an arc length that is too long will produce a great amount of spatter and flat, wide beads.

T F **6.** When welding, a humming noise usually indicates that the amperage and arc length are correct.

T F **7.** Amperage must be correct to ensure proper penetration and necessary arc control.

T F **8.** Welding with an arc length that is too short will produce too much penetration and burn-through in the base metal.

T F **9.** The type of current used, position of the weld, and properties of the base metal must be considered when selecting the correct electrode.

T F **10.** The tapping method of striking an arc is easier to learn than the scratching method.

Multiple Choice

__________ **1.** __________ is the distance between the electrode and the work.

A. Electrode gap
B. Current length
C. Travel angle
D. Arc length
E. none of the above

__________ **2.** An electrode angled slightly toward the end of the weld forms a(n) ________.

A. arc distance
B. humming noise
C. travel angle
D. work angle
E. none of the above

__________ **3.** The rate at which the electrode is moved across the weld area is the __________ __________.

A. electrode angle
B. speed of travel
C. scratching speed
D. arc length
E. none of the above

__________ **4.** If no slag is mixed in with the puddle, the __________ will appear clear and bright.

A. electrode
B. arc
C. molten metal
D. bead
E. none of the above

_______ **5.** The arc gets hotter and the _______ melts faster with increased amperage on the welding machine.

A. electrode
B. base metal
C. weld metal
D. all of the above
E. none of the above

_______ **6.** The final adjustment of _______ is made after beginning the welding operation.

A. voltage
B. amperage
C. arc blow
D. spatter
E. none of the above

_______ **7.** As the amperage is increased, _______ is greater.

A. heat
B. penetration
C. electrode burn-off
D. all of the above
E. none of the above

_______ **8.** Before starting the welding operation, _______.

A. determine the correct polarity
B. make sure the bench top is clean and dry
C. inspect cable connections
D. adjust the machine to the approximate amperage
E. all of the above

_______ **9.** What is the meaning of the welding symbol shown?

A. Butt weld arrow side
B. Bevel-groove weld arrow side
C. Flare-bevel-groove weld arrow side
D. Flare-V-groove arrow side
E. none of the above

TECH-CHEK ✓ 15

Name ______________________

Class ____________ Date ____________

Running Continuous Beads

Score ____________

True-False

T F **1.** Undercutting can occur if the amperage is set too high.

T F **2.** Padding is an operation which builds up the surface of a plate by running successive passes on the same piece.

T F **3.** Weaving an electrode is a technique used to control the heat of the arc.

T F **4.** The arc length used should be approximately the diameter of the electrode.

T F **5.** Overlapping results from welding with excessive amperage.

T F **6.** Slag should be chipped away from the body when cleaning a weld.

T F **7.** A crater is formed when the arc comes in contact with the base metal.

T F **8.** The work angle of an electrode is 90° when running continuous beads.

T F **9.** To remelt a crater, the arc should be struck approximately ½″ in front of the previously deposited bead.

T F **10.** A whipping motion is used when welding with an E-6013 electrode.

Multiple Choice

__________ **1.** When weaving the electrode, a __________ pattern may be used.
- A. rotary
- B. figure 8
- C. crescent
- D. all of the above
- E. none of the above

__________ **2.** When depositing a bead, the bead formation is controlled by the correct __________.
- A. arc length
- B. travel angle
- C. speed of travel
- D. all of the above
- E. none of the above

__________ **3.** The penetration of a deposited bead should equal __________ the total thickness of the weld bead.
- A. twice
- B. twice to three times
- C. one-eighth to one-fourth
- D. one-third to one-half
- E. none of the above

__________ **4.** The __________ angle is in the line of the welding and may vary from 5° to 30°.
- A. work
- B. travel
- C. current
- D. arc
- E. all of the above

______ 5. When the welding amperage is too low, ______.
   A. beads pile up on the base metal
   B. there is not enough heat to melt the base metal
   C. the beads will be irregular in shape
   D. all of the above
   E. none of the above

______ 6. Undercutting on the vertical plate of a T-joint can be avoided by ______.
   A. increasing the amperage
   B. changing the electrode angle
   C. increasing the arc length
   D. all of the above
   E. none of the above

______ 7. Slag is removed from the weld using a ______.
   A. long arc length
   B. electrode holder
   C. pair of pliers
   D. chipping hammer
   E. none of the above

______ 8. Evenly spaced ripples in the bead are formed by maintaining a consistent ______.
   A. arc length
   B. travel angle
   C. work angle
   D. speed of travel
   E. all of the above

______ 9. Welding amperage that is too high results in excessive ______.
   A. bead heighth
   B. travel speed
   C. spatter
   D. overlaps
   E. none of the above

______ 10. When padding (surfacing), ______ must be removed before depositing additional beads.
   A. arc blow
   B. weld metal
   C. slag
   D. crater
   E. none of the above

______ 11. What is the meaning of the welding symbol shown?

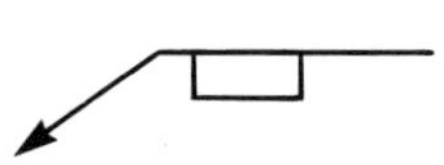

   A. Square butt joint
   B. Plug or slot weld other side
   C. Bevel groove other side
   D. Plug or slot weld arrow side
   E. none of the above

TECH-CHEK ✓ 16

Name ____________________

Class ____________ Date ____________

Flat Position

Score ____________

True-False

T F 1. The first pass of a multiple pass T-fillet joint is deposited with a work angle of 30°.

T F 2. For maximum strength, the T-joint is welded on both sides.

T F 3. The beveled butt joint is welded with a single pass.

T F 4. To weld a round shaft, it is necessary to bevel opposite sides at the same angle for maximum penetration.

T F 5. Tack welds are used to cover the filler passes deposited on the butt joint.

T F 6. The cover pass is used to provide additional reinforcement and a finished appearance.

T F 7. The welding operation is simplified if the joint is in the horizontal position.

T F 8. An open butt joint is tack welded with a $^3/_{32}$" to $^1/_8$" root opening to allow for expansion and penetration.

T F 9. When welding plates of different thicknesses, more heat should be directed to the thinner plate for proper heat control.

T F 10. Tacked pieces are placed against a firebrick to obtain the flat welding position for a lap joint.

Multiple Choice

__________ 1. The __________ produces complete penetration to the bottom surface of a butt joint.
- A. cap pass
- B. cover pass
- C. filler pass
- D. root pass
- E. none of the above

__________ 2. Edge preparation, such as beveling, is not required when welding the __________ joint.
- A. closed butt
- B. lap
- C. outside corner
- D. all of the above
- E. none of the above

__________ 3. __________ are used to maintain the proper root opening between the metal being welded, and are consumed into the weld joint.
- A. Root passes
- B. Filler passes
- C. Cap passes
- D. Tack welds
- E. none of the above

__________ 4. When welding a multiple-pass fillet lap weld, the last pass is deposited with a __________.
- A. 90° travel angle
- B. filler pass
- C. weaving motion
- D. small diameter electrode
- E. none of the above

______ **5.** When welding is done in the flat position, ______.
A. better penetration can be secured
B. the molten metal has less tendency to run
C. welding speed can be increased
D. all of the above
E. none of the above

______ **6.** The ______ pass is the first pass deposited on a multiple-pass joint.
A. filler
B. open
C. cover
D. root
E. none of the above

______ **7.** Welding thick metals requires ______.
A. beveling the edges
B. the correct root opening between the pieces
C. edge preparation
D. all of the above
E. none of the above

______ **8.** The ______ joint does not require a fillet weld.
A. lap
B. butt
C. corner
D. T
E. none of the above

______ **9.** When the thickness of the metal exceeds ______", the edge of the butt joint should be beveled.
A. $^1/_{16}$
B. $^1/_{32}$
C. $^1/_8$
D. all of the above
E. none of the above

______ **10.** The root pass should not penetrate the bottom surface of the groove more than ______".
A. $^1/_4$
B. $^5/_{16}$
C. $^3/_8$
D. all of the above
E. none of the above

______ **11.** What is the meaning of the welding symbol shown?

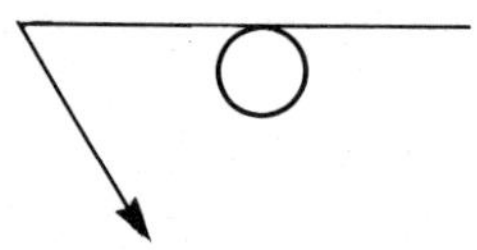

A. Plug weld arrow side
B. Spot weld
C. Butt weld
D. Bevel-groove
E. none of the above

TECH-CHEK ✓ 17

Name ______________________

Class ____________ Date ____________

Horizontal Position

Score ____________

True-False

T F **1.** A fast-freeze electrode is best suited for horizontal welding.

T F **2.** The root pass is the last pass of a multiple pass weld joint.

T F **3.** Undercut is caused by using too much amperage when welding in the horizontal position.

T F **4.** A weaving motion is used when depositing a cover pass.

T F **5.** An overlap occurs when weld metal is not properly fused to the base metal.

Multiple Choice

________ **1.** A work angle of ________° is recommended when depositing the first pass of a multiple-pass T-fillet.
- A. 10
- B. 20
- C. 45
- D. 30
- E. 70

________ **2.** A work angle of ________° is recommended when depositing the second pass of a multiple-pass T-fillet.
- A. 10
- B. 20
- C. 45
- D. 30
- E. 70

________ **3.** A work angle of ________° is recommended when depositing the third pass of a multiple-pass T-fillet.
- A. 10
- B. 20
- C. 45
- D. 30
- E. 70

________ **4.** When welding a butt joint in the horizontal position, ________ in order to deposit a properly formed bead.
- A. a short arc length must be maintained
- B. the amperage must be reduced slightly (compared to the flat position)
- C. the electrode must be angled upward about 5°–10°
- D. a slight weaving motion should be used
- E. all of the above

________ **5.** The ________ type of electrode is best suited for welding in the horizontal position.
- A. fast-freeze
- B. fill-freeze
- C. fast-fill
- D. all of the above
- E. none of the above

_____ **6.** A _____ pass is used to produce a smooth finish on the multiple-pass butt joint.
- A. filler
- B. cover
- C. root
- D. tack
- E. none of the above

_____ **7.** What is the meaning of the welding symbol shown?

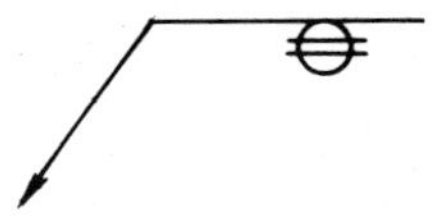

- A. Plug weld
- B. Spot weld
- C. Slot weld
- D. Seam weld
- E. none of the above

Matching

_____ **1.** A layer of metal deposited by running successive beads which overlap one another is called _____.

_____ **2.** A _____ commonly has an included angle of 60°.

_____ **3.** The recommended _____ of an electrode when depositing stringer beads in the horizontal position is 20°.

_____ **4.** The _____ of the electrode, when depositing stringer beads in the horizontal position, is 10°.

_____ **5.** A _____ is deposited on a T-joint.

- A. single V-butt joint
- B. padding
- C. travel angle
- D. work angle
- E. fillet weld

TECH-CHEK ✓ 18

Name ______________________

Class ____________ Date ____________

Score ____________

Vertical Position

True-False

T F 1. Shielded metal-arc welding vertical up is recommended for heavy plates ¼″ or greater in thickness.

T F 2. An E-6010 or E-6011 electrode is recommended for welding vertical up.

T F 3. Vertical down welding obtains greater penetration than vertical up welding.

T F 4. A whipping motion is recommended when using an E-7018 electrode for vertical welding.

T F 5. A 10°–30° travel angle is recommended for welding vertical down joints.

T F 6. Welding vertical down can be performed faster than welding vertical up.

T F 7. A filler pass is used when a smooth weld is required on the final pass of a wide joint.

T F 8. When welding a vertical up lap joint, the root pass is deposited without electrode motion.

T F 9. To ensure a strong weld, slag must be completely removed after each pass of a multiple-pass joint.

T F 10. A weaving motion is used to control the directed heat from the arc when depositing the root pass.

Multiple Choice

__________ 1. When welding a vertical down joint using an E-7018 electrode, the electrode is __________ lightly with a very short arc.
   A. whipped
   B. withdrawn
   C. dragged
   D. hesitated
   E. none of the above

__________ 2. The correct order of depositing a multiple pass fillet weld follows the sequence of __________, __________, and __________ passes.
   A. root, filler, cover
   B. filler, cover, root
   C. whip, weave, drag
   D. cover, root, weave
   E. none of the above

__________ 3. Vertical down welding is used for welding light gauge metal because the __________ is shallow.
   A. arc length
   B. whipping motion
   C. weaving
   D. penetration
   E. none of the above

__________ 4. When welding vertical down, ______________________.
   A. the arc is kept short
   B. the molten metal and slag should be kept from running ahead of the crater
   C. the electrode is pointed upward
   D. a travel angle of 15°–30° is used
   E. all of the above

_______ **5.** A _______ is used to control the weld puddle when welding vertical up.
A. drag motion
B. long arc length
C. larger diameter electrode
D. fast-fill electrode
E. none of the above

_______ **6.** A(n) _______ type of electrode is recommended for welding in the vertical position.
A. fast-freeze
B. iron powder
C. fast-fill
D. all of the above
E. none of the above

_______ **7.** When using a _______ motion, the electrode is moved out of the puddle without breaking the arc.
A. whipping
B. weaving
C. dragging
D. all of the above
E. none of the above

_______ **8.** The arc is struck at the _______ of the joint when welding vertical up.
A. top
B. bottom
C. middle
D. all of the above
E. none of the above

_______ **9.** What is the meaning of the welding symbol shown?

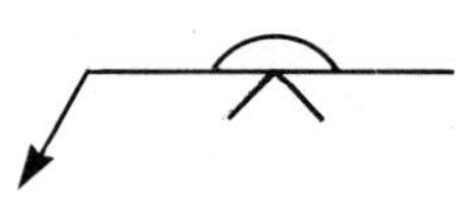

A. Groove weld with backing
B. Fillet weld
C. Groove weld with slot
D. Bevel-groove with backing
E. none of the above

TECH-CHEK ✓ 19

Name ______________________

Class ____________ Date ____________

Score ____________

Overhead Position

True-False

T F **1.** A travel angle of 15° is recommended when welding a lap joint in the overhead position.

T F **2.** The electrode holder is held with knuckles facing upward and the palm facing downward.

T F **3.** A long arc length is used to prevent molten metal from falling out of the puddle.

T F **4.** A welding cap is recommended when welding in overhead position.

T F **5.** Standing to one side of the work is advisable to avoid injury when welding in the overhead position.

Multiple Choice

_________ **1.** Of the four welding positions, _________ is the easiest to master and _________ is the most difficult.
A. horizontal, vertical
B. flat, vertical
C. overhead, flat
D. flat, overhead
E. none of the above

_________ **2.** A(n) _________ is used to secure the weld pieces when practicing welds in the overhead position.
A. vise
B. ceiling pole
C. positioner
D. arc holder
E. all of the above

_________ **3.** _________ electrodes are recommended for welding in the overhead position.
A. Fast-freeze
B. Fast-fill
C. Fill-freeze
D. Iron-powder
E. none of the above

_________ **4.** The _________ should be reduced when welding in the overhead position.
A. cable length
B. tendency for molten metal to run
C. amperage
D. travel angle
E. work angle

_________ **5.** To reduce the cable weight when welding in the overhead position, the cable should be draped over your _________ when standing and over your _________ when sitting.
A. helmet, shoulder
B. shoulder, knee
C. foot, shoulder
D. knee, shoulder
E. none of the above

______ **6.** Hold the electrode at a ______° work angle when running beads in the overhead position.
A. 45
B. 75
C. 30
D. 90
E. none of the above

______ **7.** A ______° travel angle is used when welding a lap joint in the overhead position.
A. 30
B. 10–15
C. 45
D. 90
E. none of the above

______ **8.** What is the meaning of the welding symbol shown?

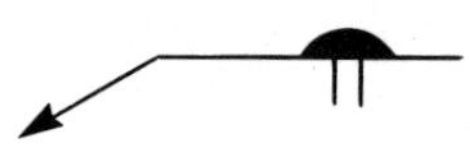

A. Square butt with melt thru
B. Bevel-groove with no joint preparation
C. Flare-groove with melt-thru
D. Double-groove weld
E. none of the above

TECH-CHEK ✓ 20

Name ______________________

Class ____________ Date ____________

Score ____________

Cast Iron

True-False

T F 1. A temperature stick can be used to determine when the correct preheat temperature has been reached.

T F 2. Keeping the casting heated to 2,000°F during the entire weld operation prevents cracks from forming in the casting.

T F 3. The casting must be thoroughly cleaned at the weld area to ensure a sound weld.

T F 4. Welding on cast iron requires a higher amperage than that for mild steel.

T F 5. Direct current straight polarity is often used when welding cast iron.

T F 6. The backstep welding technique minimizes the heat input of the weld.

T F 7. Large diameter electrodes deposit metal more quickly and are used to minimize heat when welding on cast iron.

T F 8. The weld is peened to relieve stresses caused by the welding operation.

T F 9. To obtain the necessary penetration when repairing a crack in the weld area, V the crack approximately 1/8″ to 3/16″ deep.

T F 10. On thinner castings, studs less than 1/2″ thick may be used to reinforce the weld.

Multiple Choice

__________ 1. Soft and ductile weld deposits are made with __________ type electrodes.
A. fast-freeze
B. machinable
C. nonmachinable
D. all of the above
E. none of the above

__________ 2. __________ is often used for machine castings and results from high silicon content and slow cooling of the iron.
A. Malleable cast iron
B. White cast iron
C. Gray cast iron
D. Alloy cast iron
E. all of the above

__________ 3. Cast iron should never be preheated beyond __________ in order to keep the properties of the cast iron from changing.
A. a dull red
B. 1,200°F
C. 650°C
D. all of the above
E. none of the above

__________ 4. A __________ electrode produces weld deposits that are hard and waterproof.
A. fast-freeze
B. machinable
C. nonmachinable
D. all of the above
E. none of the above

________ **5.** ____________ to prevent cracks from forming on the casting.
A. Preheat to the correct temperature
B. Peen the weld bead with a hammer after welding
C. Allow the weld piece to cool slowly
D. Run short beads approximately 1″–3″ long
E. all of the above

________ **6.** Welding is not recommended on __________.
A. gray cast iron
B. malleable cast iron
C. white cast iron
D. all of the above
E. none of the above

________ **7.** __________ is rapidly cooled, leaving it very hard and difficult to machine.
A. Gray cast iron
B. Malleable cast iron
C. White cast iron
D. all of the above
E. none of the above

________ **8.** Fine hairline cracks in a casting can be made more visible by rubbing ________ over the surface.
A. lead
B. a temperature indicating crayon
C. slag
D. chalk
E. none of the above

________ **9.** A __________ at the end of a crack in a casting prevents the crack from extending during the welding operation.
A. V
B. chill block
C. hole drilled
D. crater
E. none of the above

________ **10.** ________ contains elements added to the iron to increase tensile strength, machinability, and other properties.
A. Malleable cast iron
B. Gray cast iron
C. Alloy cast iron
D. White cast iron
E. none of the above

________ **11.** What is the meaning of the welding symbol shown?

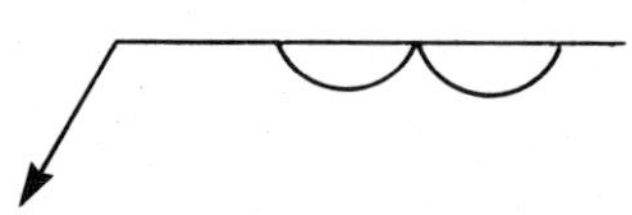

A. Back weld
B. Spot weld
C. Surfacing weld
D. Fillet weld
E. none of the above

TECH-CHEK ✓ 21

Name ______________________

Class ____________ Date ____________

Carbon Steels

Score ____________

True-False

T F **1.** Steels with a high carbon content must be preheated before they are welded.

T F **2.** E-60 series electrodes should be used when welding high-carbon steels.

T F **3.** The control heat process uses a very low heat input to minimize changes in the properties of the base metal.

T F **4.** When welding low-carbon steels, the welding heat has no appreciable effect on the parent metal.

T F **5.** Crater cracks appear on improperly formed fillet welds.

T F **6.** A 50-50 solder can be used to determine when the proper preheat temperature has been reached.

T F **7.** Plain carbon steels are made in killed, semikilled, and rimmed grades.

T F **8.** Porosity in the weld can be prevented by increasing the amperage.

T F **9.** The strength of high-carbon steel is obtained by a heat-treatment process.

T F **10.** Preheat temperatures for mild steel range from 1,200°F to 2,000°F, depending on the carbon content of the steel.

Multiple Choice

__________ **1.** __________ steel has a smooth surface and contains no blowholes.

A. Killed
B. Semikilled
C. Rimmed
D. Low-carbon
E. all of the above

__________ **2.** __________ steel is deoxidized by adding silicon or aluminum.

A. Killed
B. Semikilled
C. Rimmed
D. Alloy carbon
E. Low-carbon

__________ **3.** __________ are caused by excessive carbon that the weld bead picks up from the base metal.

A. Crater cracks
B. Porosities
C. Killed steels
D. Root cracks
E. none of the above

__________ **4.** The advantage of preheating the base metal is to __________.

A. prevent cold cracks
B. reduce distortion
C. reduce residual stresses
D. reduce hardness in heat-affected zones
E. all of the above

______ **5.** Medium-carbon steels have a carbon content of ______%.
A. 0.10–0.15
B. 0.30–0.45
C. 0.45–0.60
D. 0.60–0.75
E. none of the above

______ **6.** Using a ______ eliminates preheating because welding is done with a high current and a slow travel speed.
A. postweld heat treatment
B. porosity checker
C. control heat process
D. killed steel
E. all of the above

______ **7.** Which of the following is not a defect that might occur when welding carbon steels?
A. Crater cracks
B. Porosity
C. Root cracks
D. Process control
E. none of the above

______ **8.** ______ occur(s) when gas is entrapped in the metal, leaving small gas pockets.
A. Root cracks
B. Segregation
C. Inclusions
D. Porosity
E. none of the above

______ **9.** What is the meaning of the welding symbol shown?

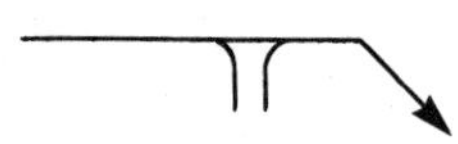

A. Flare-groove
B. Flare-bevel
C. Edge-flange
D. Single-J
E. none of the above

TECH-CHEK ✓ 22

Alloy Steels

Name ____________________

Class ____________ Date ____________

Score ____________

True-False

T F 1. A clip test may be used to determine if preheating is necessary.

T F 2. Shielded metal-arc welding on alloy steels allows welding to take place at lower temperatures, and therefore is recommended over inert-gas processes.

T F 3. Welding stainless steel requires more amperage than that required for mild steel.

T F 4. The lowest possible current should be used when welding high-manganese steel.

T F 5. E-70XX type electrodes should be used for welding low-carbon moly steels.

T F 6. Low-manganese steels are used for fabricating parts that must withstand impact stresses and resist wear.

T F 7. Chrome-moly steels with a low carbon content can be welded with E-60XX series electrodes.

T F 8. Adding nickel to steel increases its brittleness and hardness.

T F 9. Low-alloy molybdenum steel is sometimes called carbon-moly or chrome-moly steel.

T F 10. Jigs should be used to prevent distortion when stainless steel cools after welding.

Multiple Choice

________ 1. Welding stainless steel requires at least ________% less current than that required for mild steel.
   A. 10
   B. 20
   C. 30
   D. 40
   E. none of the above

________ 2. Keeping the ________ as short as possible prevents oxidation and porosity when welding stainless steel.
   A. bead
   B. electrode
   C. cladding
   D. arc length
   E. none of the above

________ 3. The 200–300 series of stainless steel requires no ________ after welding if used in normal atmospheric conditions.
   A. grain growth
   B. root cracks
   C. hardening operations
   D. annealing
   E. all of the above

________ 4. The letter ________ in the electrode classification suffix indicates the presence of chromium-molybdenum.
   A. G
   B. C
   C. B
   D. A
   E. none of the above

__________ **5.** The letter __________ in the electrode classification suffix indicates the presence of nickel steel.
A. D
B. C
C. B
D. A
E. none of the above

__________ **6.** When welding clad steel __________ are required.
A. increased amperages
B. two different electrodes
C. molybdenum steel electrodes
D. low manganese electrodes
E. none of the above

__________ **7.** The AISI (__________) classifies stainless steels.
A. Allied Identification of Steels Incorporated
B. American Iron and Steel Institute
C. American International Steel Identification
D. all of the above
E. none of the above

__________ **8.** The use of __________ helps reduce distortion from the effects of heat caused from welding.
A. chill plates
B. jigs
C. fixtures
D. all of the above
E. none of the above

__________ **9.** Alloy steels up to __________" in thickness may be welded without edge preparation.
A. 1
B. $^{1}/_{2}$
C. $^{1}/_{8}$
D. $^{1}/_{4}$
E. none of the above

__________ **10.** What is the meaning of the welding symbol shown?

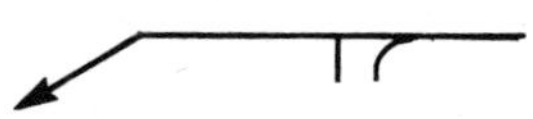

A. Flare-bevel
B. Corner-flange
C. Butt
D. Flare-groove
E. none of the above

TECH-CHEK ✓ 23

Name ____________________

Class __________ Date __________

Score __________

Nonferrous Metals

True-False

T F 1. Commercially pure aluminum does not contain any alloying ingredients.

T F 2. When welding aluminum, a short arc length should be maintained to prevent weld contamination.

T F 3. Nonferrous metals contain iron that requires special electrodes for proper penetration.

T F 4. Copper is a good conductor of heat and electricity.

T F 5. When welding brass, special precautions are required to avoid injury from toxic fumes.

T F 6. Casting alloys are used to produce aluminum castings when poured into molds.

T F 7. Aluminum less than $^{1}/_{16}$" should not be welded using the shielded metal-arc process.

T F 8. Reverse polarity should be used when welding aluminum.

T F 9. Wrought alloys of aluminum contain one or more alloying elements and possess a much higher tensile strength than pure aluminum.

T F 10. Monel and inconel obtain their strength through heat treating.

Multiple Choice

________ 1. Aluminum weighs approximately ________ less than other commonly used metals.
- A. one-fourth
- B. one-half
- C. three-fourths
- D. one-third
- E. none of the above

________ 2. ________ is a copper-tin alloy.
- A. Brass
- B. Zinc
- C. Aluminum
- D. Inconel
- E. none of the above

________ 3. ________ is a copper-zinc alloy.
- A. Monel
- B. Wrought aluminum
- C. Bronze
- D. Brass
- E. none of the above

________ 4. ________ is a common alloying element in aluminum.
- A. Manganese
- B. Silicon
- C. Zinc
- D. Copper
- E. all of the above

_______ **5.** Edge preparation is not necessary when welding aluminum _______" or less in thickness.

A. ¼
B. ½
C. ¾
D. all of the above
E. none of the above

_______ **6.** The basic temperature designation _______ indicates that the aluminum was thermally treated to produce stable tempers.

A. W
B. H
C. O
D. T
E. all of the above

_______ **7.** Aluminum in its pure state _______.

A. does not possess a high tensile strength
B. possesses high ductility
C. has a lower melting temperature than steel
D. can be welded using the shielded metal-arc process
E. all of the above

_______ **8.** What is the meaning of the dimensions on the welding symbol shown?

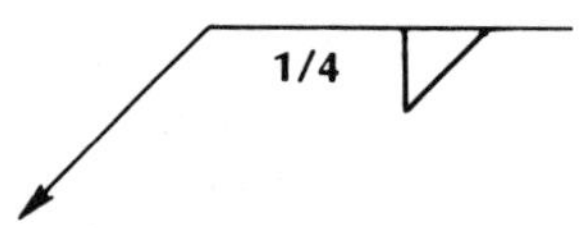

A. Unequal fillet legs
B. Flare-groove ¼" in length
C. ¼" Fillet leg size
D. ¼" Fillet weld length
E. none of the above

TECH-CHEK ✓ 24

Name ______________________

Class ____________ Date ____________

Score ____________

Gas Tungsten-Arc Welding—GTAW

True-False

T F **1.** The gas tungsten-arc welding process uses a nonconsumable tungsten electrode to provide the arc.

T F **2.** The gas shielded-arc welding process simplifies the welding process by eliminating the need for flux.

T F **3.** Gas metal-arc welding is sometimes called MIG welding.

T F **4.** Argon is often used as a shielding gas in the GTAW process.

T F **5.** The amount of electrode extension (stickout) is determined by the type of weld joint and the position of the weld.

T F **6.** Push angle refers to the torch pointing toward the beginning of the weld as the bead is deposited.

T F **7.** DCSP is often used when welding mild steel.

T F **8.** The electrode touches the work to start the arc in DC welding.

T F **9.** The filler metal used should possess the same properties as the base metal.

T F **10.** Shielding gas is not required after the arc is stopped when the weld is complete.

T F **11.** Mild steel has a lower melting point than aluminum.

T F **12.** The filler rod should be dragged in the weld puddle to increase the weld metal deposited.

T F **13.** DCRP is often used for welding mild steel.

T F **14.** Water-cooled torches are used for welding at amperages of 200 amps and above.

T F **15.** When using ACHF on aluminum, the electrode touches the work when starting the arc.

T F **16.** A flowmeter regulates the amount of shielding gas flowing to the torch.

T F **17.** A gas cup directs the shielding gas from the torch to the weld area.

T F **18.** In the semiautomatic process, welding is performed without the control of an operator.

T F **19.** Nitrogen in the air causes contamination problems in molten weld metal.

T F **20.** An electrode tip is shaped according to the type of current used in the welding operation.

Multiple Choice

________ **1.** When welding aluminum using the GTAW process, ________ is used.

A. DCRP
B. DCSP
C. AC
D. ACHF
E. none of the above

________ **2.** Of the shielding gases used for gas tungsten-arc welding, ________ allows the deepest penetration.

A. carbon dioxide
B. helium
C. argon
D. nitrogen
E. none of the above

_____ **3.** In the _____ process of application, the weld operation is done by hand.
A. automatic
B. semiautomatic
C. machine
D. manual
E. none of the above

_____ **4.** High frequency is used when welding aluminum with _____.
A. AC
B. DC
C. DCRP
D. none of the above
E. all of the above

_____ **5.** A welder can use a _____ to adjust the amount of current flowing to the weld while depositing the weld.
A. gas lens
B. flowmeter
C. ammeter
D. foot control
E. none of the above

_____ **6.** _____ provides deeper penetration than _____ when gas tungsten-arc welding.
A. AC, DC
B. DCRP, DCSP
C. DCSP, DCRP
D. ACHF, AC
E. all of the above

_____ **7.** When welding using ACHF, DCRP provides good _____, and DCSP provides good _____.
A. cleansing action, penetration
B. penetration, cleansing action
C. arc starting, cleansing action
D. penetration, arc starting
E. all of the above

_____ **8.** _____ is the amount of time the shielding gas continues to flow after the arc is stopped.
A. Flowmeter
B. Post purge
C. Foot control
D. Burnback
E. none of the above

_____ **9.** A _____ end on the electrode is used when welding with alternating current.
A. pointed
B. spherical
C. butt
D. square
E. none of the above

_____ **10.** What is the meaning of the dimension on the welding symbol shown?

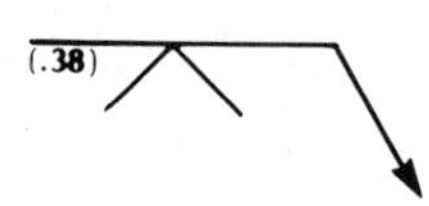

A. Bevel angle
B. Leg size
C. Effective throat size
D. Back weld size
E. none of the above

TECH-CHEK ✓ 25

Name ____________________

Class ____________ Date ____________

Score ____________

Gas Metal-Arc Welding—GMAW

True-False

T F **1.** GMAW is a faster welding process than SMAW.

T F **2.** Slag is removed more easily in the GMAW process.

T F **3.** Carbon dioxide is commonly used as a shielding gas for GMAW.

T F **4.** A pull-type welding gun is used for welding with soft and small diameter electrodes.

T F **5.** A drag angle results in more penetration than a push angle.

T F **6.** Argon is frequently used when welding aluminum with the GMAW process.

T F **7.** DCSP is used when welding mild steel with the GMAW process.

T F **8.** Globular transfer is used where low heat input is desired.

T F **9.** Electrode extension is another term for stickout.

T F **10.** Cold lap is caused by the weld metal not fusing completely with the base metal.

T F **11.** Argon can be mixed with carbon dioxide for certain shielding gas applications.

T F **12.** Short-circuit transfer is the most common type of transfer used.

T F **13.** The welding gun is removed immediately after welding is complete.

T F **14.** Hydrogen is used as a shielding gas when deep penetration is required.

T F **15.** To increase welding amperage, the wire feed speed must be increased.

T F **16.** Gas drift causes improper shielding of the weld area, resulting in weld defects.

T F **17.** Electromagnetic pinch helps advance the wire from the wire feeder in the GMAW process.

T F **18.** One factor determining the amount of gas flow required is thickness of the metal being welded.

T F **19.** A 90° work angle is used when depositing beads in the flat position.

T F **20.** The push angle used in the GMAW process requires that the welding gun be pointed toward the end of the weld.

T F **21.** The type of welding gun used depends on the type of joint being welded.

T F **22.** Rotating the current adjustment control increases the amperage on the GMAW power source.

T F **23.** The spray arc welding process is used when welding aluminum by the GMAW process.

T F **24.** In most GMAW operations, the electrode is positive and the work is negative.

T F **25.** Whiskers are caused by pushing the electrode wire through the weld puddle with too much amperage.

Multiple Choice

__________ **1.** Greater penetration is obtained using the __________ technique.
- A. whipping
- B. push
- C. drag
- D. weave
- E. none of the above

_______ 2. The _______ delivers the wire, shielding gas, and welding current to the weld area.
A. wire feeder
B. welding gun
C. power supply
D. manifold system
E. all of the above

_______ 3. _______ transfer is used when minimal heat input to the base metal is required.
A. Spray
B. Short-circuit
C. Gobular
D. Short-arc
E. none of the above

_______ 4. _______ is the shielding gas most commonly used for GMAW.
A. Oxygen
B. Hydrogen
C. Nitrogen
D. Helium
E. none of the above

_______ 5. In the GMAW process, _______ is adjusted by increasing or decreasing the wire feed speed.
A. amperage
B. voltage
C. burn-through
D. polarity
E. all of the above

_______ 6. _______ occurs when welding in windy conditions, which results in inadequate shielding of the weld area.
A. Flow-over
B. Cold lap
C. Burn-through
D. Gas drift
E. none of the above

_______ 7. The shielding gas is set at _______ cfm when running beads on mild steel using the GMAW process.
A. 60
B. 10
C. 20
D. 75
E. none of the above

_______ 8. For most welding applications, the wire stickout should be _______″.
A. $^1/_{16}$
B. $^3/_8$–$^3/_4$
C. $^1/_4$–$^3/_8$
D. $^3/_{16}$–$^1/_4$
E. none of the above

_______ 9. When using a constant potential welding machine, the _______ changes with a change in the arc length.
A. amperage
B. voltage
C. gas flow
D. all of the above
E. none of the above

TECH-CHEK ✓ 26

Name ______________________

Class ____________ Date ____________

Score ____________

Related Gas Metal-Arc Welding Processes

True-False

T F **1.** In the gas metal-arc pulsed arc welding process, current changes from DC reverse polarity to DC straight polarity each cycle per minute.

T F **2.** The globular transfer range of the GMAW pulsed-arc process requires high levels of current.

T F **3.** Buried-arc CO_2 welding uses the same standard equipment that gas metal-arc welding uses.

T F **4.** Flux-cored arc welding most commonly uses carbon dioxide as a shielding gas.

T F **5.** Submerged-arc welding uses granular flux to shield the weld area from atmospheric contamination.

T F **6.** Some flux-cored electrodes can be used without a shielding gas.

T F **7.** Flux-cored arc welding most commonly uses alternating current for flat position welding.

T F **8.** Welding amperage is adjusted by increasing or decreasing the wire feed speed at the wire feeder when flux-cored arc welding.

T F **9.** The FCAW-S process requires a shielding gas.

T F **10.** Spray transfer occurs when using the gas metal-arc—pulsed-arc process.

T F **11.** The buried-arc CO_2 process provides deep penetration with fast deposition rates.

Multiple Choice

__________ **1.** Unlike gas metal-arc welding, flux-cored arc welding uses a __________.
- A. wire feeder
- B. power source
- C. tubular wire electrode
- D. welding gun
- E. all of the above

__________ **2.** A __________ angle is often used when running beads on mild steel.
- A. push
- B. drag
- C. 45° work
- D. 30° work
- E. none of the above

__________ **3.** A gaseous shield created by the wire electrode protects the weld area in the __________ welding process.
- A. gas metal-arc
- B. submerged-arc
- C. buried-arc CO_2
- D. self-shielded flux-cored
- E. all of the above

__________ **4.** __________ starts when the trigger is pulled on the flux-cored arc welding gun.
- A. Wire feed
- B. Shielding gas flow
- C. Current flow
- D. all of the above
- E. none of the above

______ **5.** Flux-cored arc welding most commonly uses ______ current when welding mild steel.
- A. DCRP (DCEP)
- B. DCSP (DCEN)
- C. AC
- D. ACHF
- E. all of the above

______ **6.** The ______ process has its greatest applications on thicker metals.
- A. GMAW
- B. GMAW-pulsed-arc
- C. submerged-arc
- D. oxyacetylene
- E. none of the above

______ **7.** What is the meaning of the combined welding symbols shown?

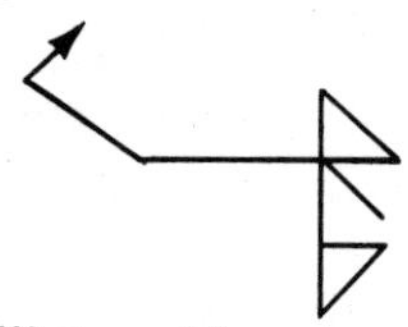

- A. J-groove with fillet welds
- B. Flare-groove with fillet welds
- C. Bevel-groove with fillet welds both sides
- D. V-groove with fillet weld arrow side
- E. none of the above

______ **8.** What is the meaning of the information in the reference tail shown?

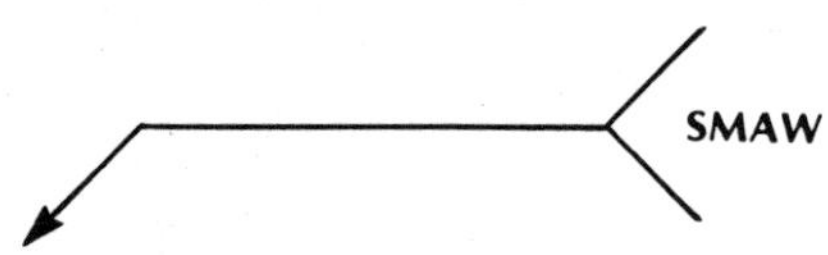

- A. Spot weld
- B. Submerged-arc welding
- C. Spray arc
- D. Shielded metal-arc welding
- E. none of the above

TECH-CHEK ✓ 27

Name ____________________

Class ____________ Date ____________

Brazing

Score ____________

True-False

T F 1. Joints used for brazing have a great amount of surface area for the adhesive qualities of the filler metal.

T F 2. Brazing requires less heat than welding.

T F 3. A brazed joint is stronger than a welded joint.

T F 4. A clean surface is required for the filler metal to flow properly in the brazed joint.

T F 5. Corrosion occurs on joints that have been brazed without the flux removed.

T F 6. The butt joint is prepared for brazing in different ways to increase the surface area of the joint.

T F 7. Brazing is done at approximately 3,200°F on mild steel.

T F 8. Braze welding is the same as brazing except that different filler metal is used.

T F 9. Brazed beads that have been overheated will not have uniform and consistent bead ripples.

T F 10. Braze welding uses capillary action to bond the weld pieces together.

T F 11. Dirt, oil, and grease must be removed before brazing is done.

T F 12. Soldering provides a stronger bond than brazing.

T F 13. Soldering requires less heat than brazing.

T F 14. The welding torch is rotated in a circular motion when braze welding.

T F 15. The point of the soldering copper should be coated with a thin layer of solder to obtain a good joint.

T F 16. When brazing, the base metal is melted to form a good joint.

T F 17. Brazing can be used to join dissimilar metals.

T F 18. Some brazing rods are coated with flux, eliminating the need for dipping as the brazing operation progresses.

T F 19. Production brazing often uses a furnace as a heat source.

T F 20. If the metal to be brazed must be preheated, the temperature should not exceed black heat.

Multiple Choice

________ 1. Braze welding should be done in the ________ position.

A. flat
B. horizontal
C. vertical
D. overhead
E. none of the above

________ 2. Brazing is performed at an approximate temperature of greater than ________°F and less than the melting point of the base metal.

A. 551
B. 800
C. 3,200
D. 6,300
E. none of the above

________ **3.** In the ________ brazing process, heat is generated in the same manner as in spot welding.
A. dip
B. manual heat
C. induction
D. furnace
E. none of the above

________ **4.** Which of the following is not a procedure used in braze welding as compared to brazing?
A. Filler metal is used.
B. The proper flux is required.
C. Heat input is kept to a minimum.
D. Filler metal is distributed by capillary action.
E. all of the above

________ **5.** Solder containing 50% lead and 50% tin melts at approximately ________°F.
A. 370
B. 800
C. 1,200
D. 471
E. none of the above

________ **6.** ________ uses silver for filler metal when joining metals.
A. Dip brazing
B. Resistance brazing
C. Induction brazing
D. Hard soldering
E. Soft soldering

________ **7.** ________ soldering is a process in which two surfaces are soldered together without allowing the solder to be seen.
A. Hard soldering
B. Seam soldering
C. Sweat soldering
D. Liquidus joint preparation
E. all of the above

________ **8.** A ________ is used to solder electronic components.
A. soldering copper
B. soldering pencil
C. flame-burning device
D. welding torch
E. none of the above

________ **9.** ________ is used to clean the soldering copper.
A. Acid
B. Sal-ammoniac
C. Tin
D. Sweat filler metal
E. none of the above

________ **10.** A(n) ________ torch can be used to provide the necessary heat for soldering.
A. air-acetylene
B. natural gas
C. mapp gas
D. propane
E. all of the above

TECH-CHEK ✓ 28

Name ______________________________

Class ____________ Date ______________

Score ____________

Surfacing

True-False

T F **1.** Hardfacing is most frequently performed in the flat position to obtain the best results.

T F **2.** Slag should be removed completely between each pass deposited on the piece using the shielded metal-arc process.

T F **3.** Hardfacing should be done with the highest possible temperature to ensure proper penetration.

T F **4.** The manufacturer should be consulted for the correct electrode to use in specific surfacing operations.

T F **5.** When using the shielded metal-arc process to surface a part, use a short arc length.

T F **6.** When using the metallizing process, the surface must be prepared by roughing to ensure a good bond of surfacing metals.

T F **7.** In any surfacing operation, the surface must be free of dirt, grease, or other foreign matter.

T F **8.** Hardfacing involves the spray coating of finely divided particles to build up worn away surfaces.

T F **9.** Chromium carbide and tungsten carbide electrodes are used to hardface surfaces for severe abrasion-resistance.

T F **10.** When hardfacing medium and high-carbon steels, preheating and postweld heat treatment are recommended.

Multiple Choice

__________ **1.** The __________ is used to measure the hardness of a particular metal.

A. clip test
B. tensile test
C. Rockwell Hardness tester
D. guided bend tester
E. all of the above

__________ **2.** __________ is the type of wear associated with grinding, rubbing, or gouging actions.

A. Corrosion
B. Impact
C. Abrasion
D. Mechanical
E. none of the above

__________ **3.** __________ is the type of wear involving the destruction of a surface from chemicals, oxidation, or scaling at elevated temperatures.

A. Corrosion
B. Impact
C. Abrasion
D. Mechanical
E. none of the above

_______ **4.** _______ is the type of wear involving crushing forces which cause parts to chip or crack.
A. corrosion
B. impact
C. abrasion
D. mechanical
E. none of the above

_______ **5.** _______ electrodes are often used for hardfacing parts that must resist impact forces without cracking.
A. High-carbon
B. Mild steel
C. Nonmachinable
D. Stainless steel
E. none of the above

_______ **6.** In _______ surfacing, an argon stream carries the metal powder surfacing material from a hopper to the electrode holder.
A. submerged-arc
B. shielded metal-arc
C. oxyacetylene
D. plasma-arc
E. all of the above

_______ **7.** In the metal spraying operation of metallizing, each coating applied has a thickness of approximately _______".
A. .125
B. .03–.05
C. .045–.060
D. .003–.005
E. none of the above

_______ **8.** The _______ hardfacing process is used when heavy deposits are required on large areas.
A. shielded metal-arc
B. oxyacetylene
C. submerged-arc
D. plasma-arc
E. none of the above

_______ **9.** A _______ motion is used to prevent burn-through when hardfacing a thin section.
A. weaving
B. whipping
C. up-and-down
D. crescent
E. all of the above

_______ **10.** When hardfacing with the oxyacetylene technique, _______.
A. the surface should be preheated
B. a slightly reducing flame is recommended
C. the tip of the hardfacing rod is held on the fringe of the flame
D. a forehand welding technique is used
E. all of the above

_______ **11.** What is the meaning of the welding symbol shown?
A. Weld around fillet arrow side
B. Weld around bevel-groove arrow side
C. Weld around V-groove arrow side
D. Weld around fillet other side
E. none of the above

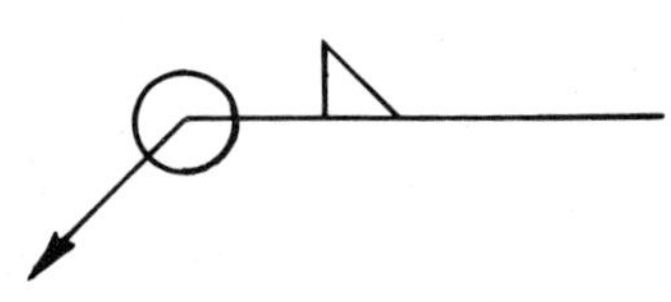

TECH-CHEK ✓ 29

Name ______________________

Class ____________ Date ____________

Score ____________

Pipe Welding

True-False

T F **1.** Heavy-wall pipe is welded using the downhill technique.

T F **2.** Tack welds are used to maintain correct position of the pipe before the first bead is deposited.

T F **3.** Pipe with wall thicknesses less than $^1/_8$″ are not beveled.

T F **4.** The uphill technique is used to obtain greater penetration than the downhill technique.

T F **5.** A root bead is deposited in the weld joint before the filler passes.

T F **6.** In the 1G test position, the pipe is rotated to maintain a flat position during welding.

T F **7.** Certification standards vary for different welding jobs and are often modified by local specifications.

T F **8.** Pipe welding is recognized as a specialized trade in itself.

T F **9.** All positions of welding are needed to weld each section of pipe when using the stove pipe method.

T F **10.** The included angle frequently used on thick wall pipe is 60°.

T F **11.** Tack welding is usually performed at four locations on the pipe joint.

T F **12.** The stringer pass (bead) requires complete penetration through the wall of the pipe.

T F **13.** A keyhole made from striking the arc is left in the joint.

T F **14.** A cap (cover) pass is deposited to provide reinforcement to the weld.

T F **15.** The bevel angle required for shielded metal-arc welding of pipe is the same as the bevel angle required for automatic gas metal-arc welding.

Multiple Choice

__________ **1.** Pipe with a wall thickness greater than __________″ is classified as thick-wall pipe.
- A. $^1/_8$
- B. $^1/_4$
- C. $^5/_{16}$
- D. $^3/_8$
- E. $^1/_2$

__________ **2.** Tack welds are made at __________° intervals around the pipe.
- A. 75
- B. 60
- C. 90
- D. 120
- E. none of the above

__________ **3.** Large diameter pipe is often joined with __________ welding equipment.
- A. shielded metal-arc
- B. automatic gas metal-arc
- C. manual gas tungsten-arc
- D. oxyacetylene
- E. none of the above

__________ **4.** The electrode is held at a __________° travel angle in the downhill technique.
A. 20–25
B. 10–15
C. 30–45
D. 45–60
C. none of the above

__________ **5.** The __________ pass is deposited after the tack welds have been made.
A. root
B. filler
C. hot
D. cover
E. none of the above

__________ **6.** The __________ pass is usually deposited with a side-to-side or a horseshoe weaving motion.
A. root
B. filler
C. hot
D. cover
E. none of the above

__________ **7.** The __________ pass burns out any slag particles that may be left from the the root bead.
A. root
B. filler
C. hot
D. cover
E. none of the above

__________ **8.** In the __________ test position, the axis of the pipe is at 45° and the pipe is not turned while welding.
A. 1G
B. 2G
C. 3G
D. 4G
E. none of the above

__________ **9.** Tack welds on pipe joints are approximately __________" long.
A. 1/2
B. 1/8
C. 3/4
D. 1/16
E. none of the above

__________ **10.** The __________ method of welding involves welding short sections of pipe, then welding the large pipe section in line with the connecting pipe.
A. position
B. stove pipe
C. roll welding
D. machine
E. all of the above

__________ **11.** What is the meaning of the welding symbol shown?

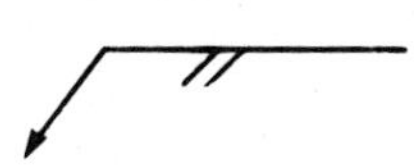

A. Butt joint arrow side
B. Butt joint other side
C. Scarf for brazed joint other side
D. Scarf for brazed joint arrow side
E. none of the above

TECH-CHEK ✓ 30

Name ______________________

Class ______________ Date ______________

Score ______________

Cutting Operations

True-False

T F **1.** The thickness of the metal being cut determines the size of the cutting tip used.

T F **2.** The cutting flame and the preheat flame should be adjusted to neutral for the best results.

T F **3.** The air carbon-arc cutting process uses compressed air at approximately 40–80 psi.

T F **4.** The air jet orifices on the air carbon-arc cutting torch should be positioned over the electrode for maximum air blast.

T F **5.** The largest hole of the cutting tip is for oxygen.

T F **6.** The gas pressures used for cutting are determined by the size of the tip.

T F **7.** In making a bevel cut, the cutting torch is held at an angle to the metal.

T F **8.** Needle valves on the cutting torch are adjusted to obtain a neutral preheat flame.

T F **9.** The oxygen control lever activates the flow of oxygen at a pressure higher than that used for oxyacetylene welding.

T F **10.** A straight, neat cut is obtained by holding the cutting torch steady throughout the entire cut.

Multiple Choice

__________ **1.** The air carbon-arc cutting process uses __________ or __________ current.
A. DCRP, alternating
B. DCSP, direct
C. alternating, direct
D. ACHF, direct
E. none of the above

__________ **2.** Plasma-arc cutting uses __________.
A. DCRP
B. DCSP
C. AC
D. ACHF
E. none of the above

__________ **3.** A __________ flame is used for cutting cast iron.
A. carburizing
B. oxidizing
C. neutral
D. all of the above
E. none of the above

__________ **4.** To ignite the flame on the cutting torch, the __________ is opened first.
A. oxygen control lever
B. oxygen needle valve
C. acetylene needle valve
D. all of the above
C. none of the above

________ **5.** Oxygen with ________ can be used for flame cutting operations.
A. acetylene
B. propane
C. mapp gas
D. natural gas
E. all of the above

________ **6.** When flame cutting round stock, the cut should be started approximately ________° from the top edge.
A. 45
B. 75
C. 90
D. 60
E. none of the above

________ **7.** A number ________ is recommended for cutting metal that is ¾″ thick.
A. 00
B. 0
C. 1
D. 2
E. 3

________ **8.** The flame cutting process is possible by rapidly ________ the metal by introducing oxygen under pressure into the molten pool.
A. covering
B. cooling
C. oxidizing
D. blasting
E. none of the above

________ **9.** The preheat holes on the cutting tip can be identified by their ________ compared to the oxygen hole.
A. small size
B. number
C. location
D. all of the above
E. none of the above

________ **10.** When gouging using the air carbon-arc torch, the electrode must point ________ when working in a vertical direction.
A. upward
B. downward
C. horizontally
D. all of the above
E. none of the above

________ **11.** What is the meaning of the welding symbol shown?

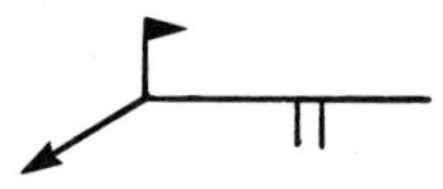

A. V-groove arrow side, weld made in field
B. V-groove other side, weld made in field
C. Butt joint arrow side, weld made in field
D. Bevel-groove arrow side
E. none of the above

TECH-CHEK ✓ 31

Name ______________________

Class ____________ Date ____________

Production Welding

Score ____________

True-False

T F **1.** Stud welding eliminates the need for drilling and tapping studs used in the manufacturing process.

T F **2.** Argon and helium are commonly used in the gas tungsten-arc spot welding process.

T F **3.** The size and configuration of the spot welder to be used depends on the welding application.

T F **4.** Spot welders use alternating current only.

T F **5.** In resistance welding, current passing through the weld pieces create the heat necessary for fusion.

T F **6.** Electron beam welding is often used for metals that are difficult to weld.

T F **7.** Laser welding permits a highly concentrated heat input in a large area.

T F **8.** Laser light beams are produced in a man-made ruby rod.

T F **9.** Gas tungsten-arc spot welding requires access to both sides of the weld pieces.

T F **10.** Flash welding and butt welding are types of resistance welding.

T F **11.** Low current and high voltage generate the heat necessary for fusion in resistance welding.

T F **12.** Closely spaced welds made by the seam welding process result in an airtight seam.

T F **13.** In resistance welding, thicker metals require shorter current time than thinner gauge materials.

T F **14.** Current flow is interrupted a number of times during the pulsation welding process.

T F **15.** Gas tungsten-arc spot welding requires good surface contact between the two weld pieces to ensure a strong weld bond.

T F **16.** When stud welding, the studs have a recess in the welding end which contains flux used to stabilize the arc and deoxidize the molten metal.

T F **17.** In the plasma-arc welding process, a gas is necessary for the plasma and the shielding of the weld area.

T F **18.** Ultrasonic welding is useful for welding very thin metals.

T F **19.** Electro-gas welding is used on metals with thicknesses from $1/8''$ to $1/4''$.

T F **20.** In inertia welding, one of the weld pieces is held stationary while the other is rotated.

Multiple Choice

________ **1.** The plasma-arc welding process uses ____________.

A. DCSP
B. DCRP
C. AC
D. ACHF
E. all of the above

________ **2.** The electron beam welding process used in the vacuum chamber produces welds without ________.
A. atmospheric contamination
B. shielding gas
C. a large amount of heat input
D. all of the above
E. none of the above

________ **3.** In ________ welding, constant pressure must be applied to the weld pieces while current is passed through.
A. flash
B. butt
C. inertia
D. laser
E. all of the above

________ **4.** Seam welding is performed with rotary ________.
A. voltage
B. projections
C. fusion
D. electrodes
E. all of the above

________ **5.** Volume and length of current are regulated by the ________ of the resistance welding machine.
A. frame
B. electrical circuit
C. timing controls
D. electrodes
E. none of the above

________ **6.** Resistance welding takes place when the weld metal is ________.
A. molten
B. in a plastic state
C. in a plasma state
D. in a cryogenic state
E. none of the above

________ **7.** What is the meaning of the dimension shown on the welding symbol?

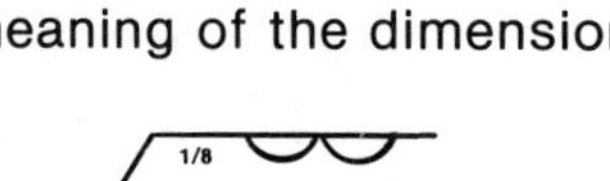

A. Scarf joint the $^1/_{16}''$ root opening
B. Butt joint with $^1/_{16}''$ root opening
C. Surface weld with $^1/_{16}''$ throat
D. Butt joint with $^1/_{16}''$ melt-thru
E. none of the above

________ **8.** What is the meaning of the dimension shown on the welding symbol?

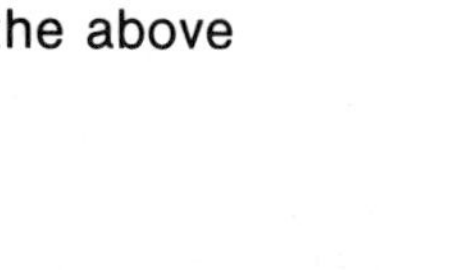

A. Surface weld $^1/_8''$ in length
B. Melt-thru $^1/_8''$
C. Surface weld $^1/_8''$ in thickness
D. Surface weld
E. none of the above

TECH-CHEK ✓ 32

Name ______________________

Class ____________ Date ____________

Score ____________

Plastic Welding

True-False

T F **1.** The edges of plastic weld pieces should be beveled to an included angle of 60°.

T F **2.** Thermosetting plastics are not weldable.

T F **3.** A permanent bond in plastic is achieved by forcing a heated filler rod into the softened surface of the joint.

T F **4.** Friction or spin welding is useful in fastening knobs, containers, or other parts that can be easily spun.

T F **5.** When using the high-speed welding tip, the filler rod should remain in the welding tip for adequate preheating.

T F **6.** In induction welding, a metallic insert must remain in the weld.

T F **7.** In the heated-tool welding process, the pieces are heated after being pressed together.

T F **8.** In the hot gas welding process, the weld beads will not separate from the base material if welded correctly.

T F **9.** In high-speed welding, insufficient welding speed results in the filler rod stretching from excessive heat.

T F **10.** Compressed air or nitrogen gas is used for hot gas welding.

T F **11.** Hot gas welding guns supply temperatures ranging from 400°–925°F.

T F **12.** Air pressure in the hot gas welding process depends on the temperature used and varies from 20 to 60 psi.

T F **13.** A fanning motion is used to maintain consistent heat on the filler rod and the weld area.

T F **14.** Rods in the roll form are used in high-speed welding.

T F **15.** Sufficient pressure should be applied to bend the filler rod when feeding it into the weld area.

Multiple Choice

__________ **1.** Soft material squeezed out beyond the weld area in the inertia welding process is called __________.

A. cold lap
B. filler stretch
C. flashing
D. filler manipulation
E. none of the above

__________ **2.** Nitrogen is used as a welding gas when welding __________.

A. polyethylene
B. polypropylene
C. ABS
D. all of the above
E. none of the above

__________ **3.** __________ to avoid injury from odors and fumes caused from plastic welding.

A. A respirator should be used when welding certain plastics
B. Proper ventilation is required
C. Manufacturer's recommendations for safe practices should be followed
D. Chemical composition of the plastic being welded must be known
E. all of the above

_______ **4.** Which of the following does not apply to welding both plastics and metals?
A. The weld joints used for both plastics and metals are the same.
B. Filler material is used to strengthen the joint.
C. The base material is in a molten state before adding filler rod.
D. Adequate ventilation is required.
E. none of the above

_______ **5.** _______ is a plastic from the thermoplastic family.
A. Polyester
B. Urethane
C. Polyvinyl chloride
D. Silicone
E. none of the above

_______ **6.** _______ is a plastic from the thermosetting family.
A. Polyester
B. Epoxy
C. Urethane
D. all of the above
E. none of the above

_______ **7.** _______ welding consists of rubbing the surfaces of the parts to be joined until sufficient heat is developed for the fusing temperature.
A. Heated tool
B. Induction
C. Flash
D. Friction
E. none of the above

_______ **8.** In the hot gas welding technique, heat is directed _______.
A. to the rod
B. to the joint
C. with a fanning motion
D. at a 45° angle
E. all of the above

_______ **9.** _______ filler rods produce faster welds on V and fillet welds.
A. Round
B. Flat
C. Trianglar
D. Bar
E. none of the above

_______ **10.** A charred or discolored weld indicates _______.
A. underheating
B. overheating
C. cold laps
D. all of the above
E. none of the above

_______ **11.** What is the meaning of the welding symbol shown?

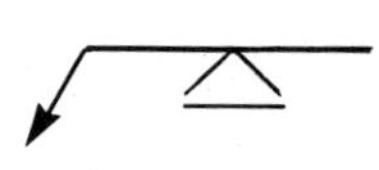

A. Bevel-groove other side, flush contour
B. V-groove arrow side, flush contour
C. Double bevel-groove other side, concave contour
D. Flange weld with convex contour
E. none of the above

TECH-CHEK ✓ 33

Name ____________________

Class ____________ Date ____________

Score ____________

Robotics and Welding

True-False

T F 1. The rectilinear robot is used to weld large pieces that do not require intricate maneuvering.

T F 2. The industrial robot must be programmed to perform the desired tasks.

T F 3. The robot welding machine, given the necessary production setting, can reduce production costs by increasing productivity.

T F 4. The working volume of a robot is its maximum range of movement capability.

T F 5. The positioner does not have to be synchronized with the movements of the robot.

T F 6. The articulating-type robot provides a flexible working volume using limited floor space.

T F 7. Robotic welding is a semiautomatic welding application used in industry today.

T F 8. With advances in programming, the use of robot welding systems is increasing.

T F 9. Robot welding machines and their supporting components are complex and expensive.

T F 10. A robot is classified by its number of axes.

Multiple Choice

__________ 1. An advantage of robotics in welding is that it __________.
- A. yields consistent quality
- B. eliminates the use of costly jigs
- C. increases productivity
- D. adapts to different weld locations
- E. all of the above

__________ 2. Robot welding systems can use the ________ process.
- A. GMAW
- B. FCAW
- C. RSW
- D. GTAW
- E. all of the above

__________ 3. The estimated arc-on time of a welding machine operator and a robot working together exceeds __________%.
- A. 30
- B. 50
- C. 60
- D. 70
- E. none of the above

__________ 4. The robot welding machine is a __________ type of weld application.
- A. manual
- B. semiautomatic
- C. machine
- D. automatic
- E. none of the above

_______ **5.** The _______ can be used to program specific weld functions by moving the robot through each operation while entering necessary data into its memory.

A. robot controller
B. positioner
C. automatic welding equipment
D. control pendant
E. all of the above

_______ **6.** What is the meaning of the welding symbol shown?

A. Bevel-groove other side, flush contour
B. V-groove arrow side, convex contour
C. Double V-groove, convex contour
D. V-groove other side, flush contour
E. none of the above

Matching

_______ **1.** The robot _______.
_______ **2.** The robot controller _______.
_______ **3.** The positioner _______.
_______ **4.** Automatic welding equipment _______.
_______ **5.** Operator controls _______.

A. translates and relays commands from the operator controls
B. locates the workpiece in a predetermined position
C. start, execute the program, and end the robot welding cycle
D. positions the welding equipment in the correct location
E. is the same as that used for other types of applications

TECH-CHEK ✓ 34

Name ______________________

Class ____________ Date ____________

Score ____________

Testing Welds

True-False

T F 1. Destructive testing subjects weld samples to loads until they fail.

T F 2. Before a specimen is placed in a tensile testing machine, an accurate measurement should be taken for calculating percent of elongation.

T F 3. Slag inclusions are always checked by visual examination.

T F 4. A tensile specimen for an all-weld metal requires threading on both ends.

T F 5. The soundness of a weld is the degree of freedom it has from defects that can be found by visual inspection.

T F 6. Tensile strength of a weld is the amount of resistance it has to bending.

T F 7. The magnetic particle inspection test requires that the weld piece be magnetized.

T F 8. The dye penetrant test is particularly useful in locating defects in a nonferrous metal, such as aluminum.

T F 9. The guided-bend and free bend tests are types of impact tests used to determine soundness.

T F 10. The face bend specimen is used to check the quality of fusion in the weld.

Multiple Choice

________ 1. A black light with a fluorescent liquid is used in the ________ inspection method.
   A. visual
   B. radiograph
   C. dye penetrant
   D. magnetic particle
   E. all of the above

________ 2. ________ testing is a dynamic test in which the weld is broken by a single blow.
   A. Etch
   B. Tensile
   C. Shear
   D. Guided-bend
   E. none of the above

________ 3. A ________ weld is tested for soundness by applying force with a press, a testing machine, or hammer blows.
   A. pipe
   B. butt tensile
   C. nick-break
   D. Charpy
   E. fillet

________ 4. The ________ testing method is the quickest way to test welds.
   A. destructive
   B. nondestructive
   C. visual
   D. all of the above
   E. none of the above

__________ 5. The __________ method determines the soundness of a weld with x-rays and gamma rays.
   A. ultrasonic testing
   B. radiographic inspection
   C. Izod
   D. magnetic particle
   E. none of the above

__________ 6. What is the meaning of the welding symbol shown?

   A. V-groove other side, concave contour
   B. V-groove arrow side, convex contour
   C. Bevel-groove, flush contour
   D. Bevel-groove both sides
   E. none of the above

Matching

Identify the following testing methods.

__________ 1. The __________ test indents the metal to determine the hardness of the metal.

__________ 2. The __________ test forces a ball into the surface of the metal.

__________ 3. The __________ test uses high frequency vibrations to locate fine surface and subsurface cracks.

__________ 4. The __________ test uses acid to make visible the boundary between the weld metal and the base metal.

__________ 5. In the __________ test high-frequency alternating current brought close to a metal produces a current in the metal by induction.

A. Brinell
B. ultrasonic
C. Rockwell hardness
D. etch
E. eddy current

TECH-CHEK ✓ 35

Name ______________________

Class ____________ Date ____________

Score ____________

Reading Welding Symbols

True-False

T F 1. A fillet is a type of weld joint used in fabrication.

T F 2. Arc welding processes include gas tungsten-arc welding and gas metal-arc welding.

T F 3. The field weld symbol is shown as a darkened triangular flag.

T F 4. Melt-thru welds occur when the weld completely penetrates through the weld piece.

T F 5. A reference tail must be used when describing all weld joints and types.

Multiple Choice

________ 1. The ________ welding process is specified on the welding symbol shown.

A. spot metallic-arc
B. shielded metal-arc
C. soldered metal-air
D. solid metal-arc
E. all of the above

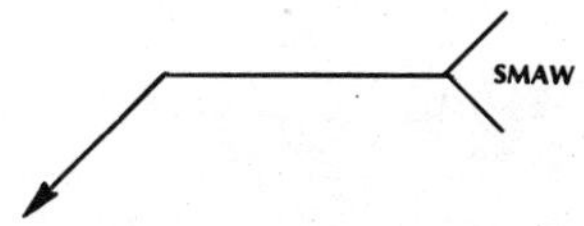

________ 2. The weld symbol shown below is a ______________________.

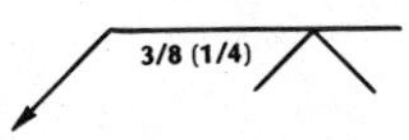

A. V-groove with $^{3}/_{8}$″ groove depth and $^{1}/_{4}$″ effective throat
B. V-groove with $^{1}/_{4}$″ groove depth and $^{3}/_{8}$″ effective throat
C. bevel-groove with $^{3}/_{8}$″ penetration spaced $^{1}/_{4}$″
D. double-bevel butt with $^{3}/_{8}$″ root opening and $^{1}/_{4}$″ groove depth
E. none of the above

________ 3. The weld symbol shown below is a ________.

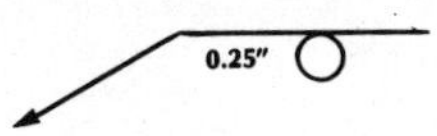

A. square butt
B. 0.25″ spot weld
C. 0.25″ bevel butt
D. 2.50″ fillet weld
E. none of the above

Matching

Identify the types of welds shown.

________ 1. Plug
________ 2. Seam
________ 3. Fillet
________ 4. V

A B C D

________ **5.** V
________ **6.** Square
________ **7.** Flare-V
________ **8.** Fillet

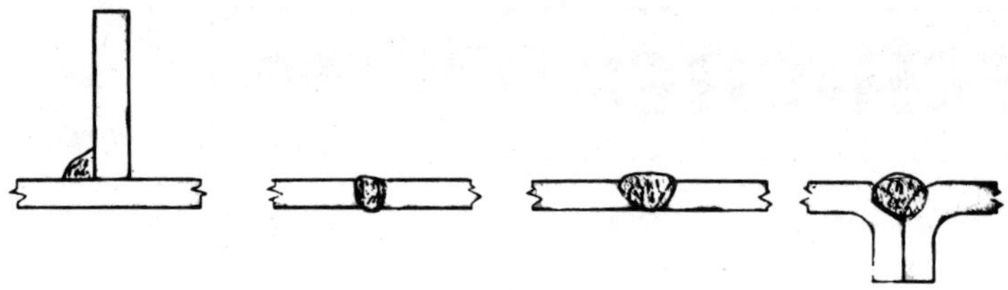

A B C D

________ **9.** Spacing
________ **10.** Number
________ **11.** Strength
________ **12.** Size

A B C D

TECH-CHEK ✓ 36

Name ______________________

Class ____________ Date ____________

Score ____________

Certification of Welders

True-False

T F **1.** In the tests for structural welding qualifying, the root bend and face bend tests are used to determine weld soundness.

T F **2.** The impact test is used to determine ductility of a weld.

T F **3.** A ¼″ electrode used for vertical and overhead positions is a qualification requirement for structural welding.

T F **4.** Specifications are specific regulations which cover the quality of a particular product to be fabricated by welding.

T F **5.** Welder certification is usually specified for a particular joint and position, and weld specification.

Multiple Choice

__________ **1.** Qualifying tests are performed on __________ of the same material used in the product involved.
- A. samples
- B. profiles
- C. certificates
- D. coupons
- E. none of the above

__________ **2.** The __________ AWS weld test position specifies flat position for groove welds.
- A. 1G
- B. 2G
- C. 3G
- D. 4G
- E. none of the above

__________ **3.** The __________ AWS weld test position specifies horizontal position for groove welds.
- A. 1G
- B. 2G
- C. 3G
- D. 4G
- E. none of the above

__________ **4.** The __________ AWS weld test position specifies vertical position for groove welds.
- A. 1G
- B. 2G
- C. 5G
- D. 6G
- E. none of the above

__________ **5.** The __________ AWS weld test position specifies overhead position for groove welds.
- A. 1G
- B. 2G
- C. 3G
- D. 4G
- E. none of the above

______ **6.** The ______ test position is not used for pipe and tubing for qualification.
A. 1G
B. 2G
C. 3G
D. 5G
E. 6G

______ **7.** Codes for welding have been established by the ______.
A. AWS
B. ASME
C. API
D. ANSI
E. all of the above

______ **8.** A ______ test is used to check the lack of fusion or cracks and proper weld contour as a certification requirement.
A. free bend
B. radiographic
C. tension
D. fillet weld
E. none of the above

______ **9.** A welder must be certified ______.
A. when applying for any new job
B. for a specific welding task
C. again if assigned to another job
D. all of the above
E. none of the above

______ **10.** The two basic destructive qualifying tests for pipe and tubing are the ______ and the ______.
A. free bend, fillet
B. root bend, face bend
C. tensile, guided bend
D. impact, tensile
E. none of the above

______ **11.** Degrees of welding competencies may be expressed in ______.
A. specifications
B. standards
C. codes
D. all of the above
E. none of the above

SECTION ACTIVITIES 1

Chapters 1-4 Introduction to Welding

1. List and define three major welding processes used in industry.
   A. ______________________________
   B. ______________________________
   C. ______________________________

2. List five principal job titles of welders.
   A. ______________________________
   B. ______________________________
   C. ______________________________
   D. ______________________________
   E. ______________________________

3. List five basic rules contributing to the safe handling of oxyacetylene welding equipment.
   A. ______________________________
   B. ______________________________
   C. ______________________________
   D. ______________________________
   E. ______________________________

4. List five methods used to clean a container before it is welded or cut.
   A. ______________________________
   B. ______________________________
   C. ______________________________
   D. ______________________________
   E. ______________________________

5. Sketch the following weld joints in the space provided.
   A. Butt B. T C. Lap D. Edge E. Corner

6. Sketch the following weld types in the space provided.
   A. Surface B. Fillet C. Groove D. Plug E. Slot

7. Sketch the following types of butt joints in the space provided.
A. Square B. Single Bevel C. Single-V D. Double-V E. Single-U F. Double-U

8. Sketch the following types of T-joints in the space provided.
A. Square B. Single Bevel C. Double Bevel D. Single-J E. Double-J

9. Sketch the following types of lap joints in the space provided.
A. Single Fillet Lap B. Double Fillet Lap

10. Sketch the following types of weld joints in the space provided.
A. Flush Corner B. Half-Open Corner C. Full-Open Corner D. Edge

11. Sketch the following welding positions in the space provided.
A. Flat B. Horizontal C. Vertical D. Overhead

12. Sketch and identify the parts of a fillet weld in the space provided.

13. Sketch and identify the parts of a groove weld in the space provided.

14. Obtain different steel samples from the instructor and determine the carbon content by using the spark test. Observe all safety precautions when using the grinder.

15. Determine the content of the three types of steel listed using the AISI Steel Code Classification System.

AISI Numbers	Significance
A. C1078	______________________
B. C1045	______________________
C. E50100	______________________

16. List five categories of chemical hazard information that must be contained on an MSDS.

____

____

____

____

____

SECTION ACTIVITIES 2

Chapters 5-11 Oxyacetylene Welding — OAW

Exercise 2-1. Lighting the Oxyacetylene Torch

Conditions:

Number 0 welding torch tip
Welding torch
Striker
Oxyacetylene welding equipment

Performance:

The welder will demonstrate the correct procedure for lighting an oxyacetylene welding torch.

Criteria:

Correct sequence of procedures as evaluated by the instructor.

Procedure:

1. Mount a number 0 tip on the welding torch. Turn the regulator adjusting screws all the way out.
2. Stand aside and open the oxygen and acetylene cylinder valves slowly. The oxygen cylinder valve should be opened all the way and the acetylene cylinder valve should be opened approximately one-fourth of a turn to one complete turn.
3. Stand to the side and turn the regulator adjusting screws to obtain the required working pressures.
4. Open the acetylene needle valve on the torch approximately one-fourth of a turn.
5. With the tip pointing down, position the striker approximately 1″ away from the end of the tip. Ignite the flame quickly to avoid wasting gas.
6. Adjust the flame so that it is jumping slightly away from the torch tip. If the flame produces a great amount of black smoke, increase the amount of acetylene by opening the acetylene needle valve.
7. With the acetylene burning, slowly open the oxygen needle valve to produce a neutral flame.
8. An excess amount of oxygen will produce an oxidizing flame. An excess amount of acetylene will produce a carburizing flame.

ACETYLENE BURNING IN ATMOSPHERE

EXCESS OF ACETYLENE

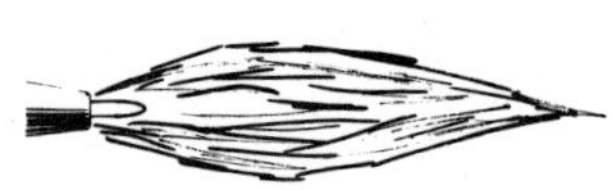

NEUTRAL FLAME

9. To shut off the torch, close the oxygen needle valve first.
10. Promptly close the acetylene valve.
11. Close the oxygen and acetylene cylinder valves to shut down the entire welding unit.
12. To remove pressure on the regulators, open the oxygen needle valve on the torch then open the acetylene needle valve. Close the needle valves after no pressure is indicated on the regulators.
13. Screw out the adjusting screws on the regulators to release pressure on the regulator.

Exercise 2-2. Testing the Flames

Conditions:

Refer to Exercise 2-1.

Performance:

The welder will demonstrate the correct procedure for adjusting the welding torch to obtain carburizing, oxidizing, and neutral flames.

Criteria:

Correct flame characteristics as evaluated by the instructor.

Procedure:

1. Refer to Exercise 2-1, Steps 1–8.

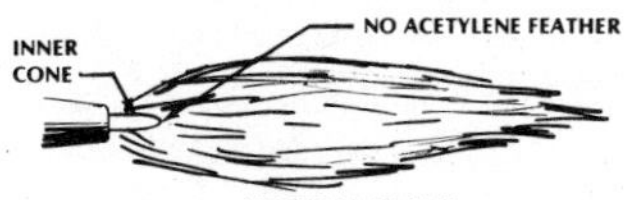

NEUTRAL FLAME

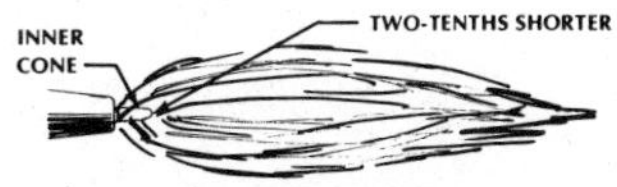

OXIDIZING FLAME

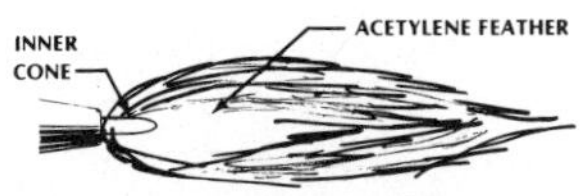

REDUCING OR CARBURIZING FLAME

2. Obtain a neutral, carburizing, and oxidizing flame by increasing or decreasing the flow of oxygen and/or acetylene.
3. Identify the characteristics of the flame.
4. Apply each flame to metal and observe the results of each.

Exercise 2-3. Carrying a Puddle Without a Filler Rod Using the OAW Process

Conditions:

Flat position
Refer to Exercise 2-1.
$^1/_{16}$″ × 3″ × 5″ mild steel

Performance:

The welder will demonstrate the correct procedure for carrying a puddle without a filler rod.

Criteria:

Weld beads should be consistent in width, ripple formation, and penetration, and run parallel to the length of the piece.

Procedure:

1. Be sure the surface of the metal is free of oil, dirt, and scale.
2. Light the torch and adjust for a neutral flame.
3. Hold the torch at a 45° angle with the inner cone of the flame $^1/_8$″ from the work.
4. Use a circular motion with the torch to distribute the heat evenly. Start from the right side of the piece, moving to the left. Reverse the direction if you are left-handed.
5. Maintain a consistent travel speed to prevent burn-throughs in the plate.
6. Practice running beads approximately $^3/_8$″ apart until properly formed beads are produced consistently.

Exercise 2-4. Laying Beads With a Filler Rod Using the OAW Process

Conditions:

Flat position
Refer to Exercise 2-2.

Performance:

The welder will demonstrate the correct procedure for laying beads with a filler rod.

Criteria:

Weld beads should be consistent in width, ripple formation, and penetration, and run parallel to the length of the piece.

Procedure:

1. Refer to Exercise 2-3, Steps 1–5.
2. Maintain a 45° angle to the work with the torch and the filler rod.
3. Dip the filler rod into the puddle with an in-and-out motion while manipulating the torch.
4. At the end of the weld, withdraw the torch slightly from the piece and fill the crater by dipping the filler rod into the puddle.
5. Practice running beads approximately 3/8″ apart until properly formed beads are produced consistently.

Exercise 2-5. Welding a Butt Joint in Flat Position Using the OAW Process

Conditions:

Flat position
Refer to Exercise 2-1.
Two pieces of 1/16″ × 1 1/2″ × 5″ mild steel

Performance:

The welder will demonstrate the correct procedure for welding a butt joint in flat position.

Criteria:

Welding technique, weld appearance, and weld strength as evaluated by the instructor.

Procedure:

1. Space the pieces of metal for progressive spacing, or tack weld the metal pieces to form a butt joint. Allow a gap of 1/16″ at the starting end of the joint.
2. Use the same motion with the torch and filler rod as practiced when running beads with filler rod. Add sufficient filler rod to build up the weld bead approximately 1/16″ above the surface of the metal.
3. Maintain a molten puddle approximately 1/4″ to 3/8″ wide.
4. Advance the torch approximately 1/16″ each rotation of the torch movement to maintain consistent bead width.
5. Fill the crater at the end of the weld to prevent a weak spot in the weld.

Exercise 2-6. Welding a Flange Joint in Flat Position Using the OAW Process

Conditions:

Flat position
Refer to Exercise 2-5.

Performance:

The welder will demonstrate the correct procedure for welding a flange joint in the flat position.

Criteria:

Welding technique, weld appearance, and weld strength as evaluated by the instructor.

Procedure:

1. Prepare the joint by bending the pieces to obtain a flange that extends above the surface of the piece approximately the thickness of the piece.
2. Butt the two flange edges and tack weld.
3. Hold the torch on the starting end until a puddle is formed.
4. Carefully manipulate the torch to maintain a consistent puddle across the joint.
5. Withdraw the torch at the end of the joint to prevent burning a hole in the joint.

Exercise 2-7. Welding a Corner Joint in Flat Position Using the OAW Process

Conditions:

Flat position
Refer to Exercise 2-5.

Performance:

The welder will demonstrate the correct procedure for welding a corner joint in flat position.

Criteria:

Welding technique, weld appearance, and weld strength as evaluated by the instructor.

Procedure:

1. Tack weld the two pieces to form a corner joint.
2. Refer to Exercise 2-6, Steps 3–5.
3. If additional build-up is required, filler rod may be added as the puddle is carried across the joint.

Exercise 2-8. Welding a Lap Joint in Flat Position Using the OAW Process

Conditions:

Flat position
Refer to Exercise 2-5.

Performance:

The welder will demonstrate the correct procedure for welding a lap joint in the flat position.

Criteria:

Welding technique, weld appearance, and weld strength as evaluated by the instructor.

Procedure:

1. Tack weld the two pieces to form a lap joint.
2. Place a firebrick under the bottom piece to obtain the flat position.
3. Use a semicircular motion with the torch. Direct more of the welding heat to the bottom piece by increasing the duration of the torch movement on the bottom piece. This will prevent overheating of the top piece.
4. Weld one side of the joint and repeat the weld on the other side.

Exercise 2-9. Welding a T-Joint in Flat Position Using the OAW Process

Conditions:

Flat position
Refer to Excercise 2-5.

Performance:

The welder will demonstrate the correct procedure for welding a T-joint in the flat position.

Criteria:

Welding technique, weld appearance, and weld strength as evaluated by the instructor.

Procedure:

1. Tack weld the two pieces to form a T-joint.
2. Place a firebrick under one side to obtain flat position.
3. Hold the torch at a 45° angle to the bottom piece.
4. Use the same technique used for welding a lap joint, but direct the heat from the torch equally to both pieces.
5. Use a semicircular torch movement to distribute the weld metal and prevent undercutting.

Exercise 2-10. Padding Using the OAW Process

Conditions:

Flat position
Refer to Exercise 2–4.

Performance:

The welder will deposit a surface weld on the metal provided.

Criteria:

The thickness of the metal should increase a minimum of $^1/_{16}$" with no voids in the buildup of weld metal.

Procedure:

1. Refer to Exercise 2-4.
2. Deposit a bead on the metal $^1/_4$" from the edge parallel to the length of the piece.
3. Deposit the next bead, penetrating half into the first bead.
4. Continue depositing beads until the plate is covered with beads with no voids in the weld metal. Quench the piece as necessary to avoid overheating.
5. Run a second layer of beads parallel to the short side of the piece.

Exercise 2-11. Welding a Butt Joint in Horizontal Position Using the OAW Process

Conditions:

Horizontal position
Refer to Exercise 2-5, Steps 2–5.

Performance:

The welder will demonstrate the correct procedure for welding a butt joint in horizontal position.

Criteria:

Welding technique, weld appearance, and weld strength as evaluated by the instructor.

Procedure:

1. Tack weld the two pieces to form a butt joint. Secure the tack welded piece in a positioner.
2. Refer to Exercise 2-5, Steps 2–5.
3. To prevent overheating, direct more heat to the bottom plate.
4. Dip the filler rod above the center of the puddle to fill the crater and to avoid undercutting.

Exercise 2-12. Welding a T-Joint in Horizontal Position Using the OAW Process

Conditions:

Horizontal position
Refer to Exercise 2-9, Steps 3–5.

Performance:

The welder will demonstrate the correct procedure for welding a T-joint in horizontal position.

Criteria:

Welding technique, weld appearance, and weld strength as evaluated by the instructor.

Procedure:

1. Tack weld the two pieces to form a T-joint.
2. Refer to Exercise 2-9, Steps 3–5.
3. Add filler rod to the vertical piece to prevent undercutting.

Exercise 2-13. Welding a Lap Joint in Horizontal Position Using the OAW Process

Conditions:

Horizontal position
Refer to Exercise 2-5.

Performance:

The welder will demonstrate the correct procedure for welding a lap joint in horizontal position.

Criteria:

Welding technique, weld appearance, and weld strength as evaluated by the instructor.

Procedure:

1. Tack weld the two pieces to form a lap joint.
2. Refer to Exercise 2-8, Steps 3–4.

Exercise 2-14. Welding a Butt Joint in Vertical Position Using the OAW Process

Conditions:

Vertical position
Refer to Exercise 2-5.

Performance:

The welder will demonstrate the correct procedure for welding a butt joint in vertical position.

Criteria:

Welding technique, weld appearance, and weld strength as evaluated by the instructor.

Procedure:

1. Tack weld the two pieces to form a butt joint. Secure the welded piece in a positioner.
2. Start the weld at the bottom of the joint.
3. Hold the torch and the filler rod at the same angle to the work as in flat position.
4. If the weld puddle becomes too large, withdraw the flame slightly. Directing the flame more toward the filler rod will also prevent a large puddle from forming, which results in undercutting and weld sag.

Exercise 2-15. Welding a T-Joint in Vertical Position Using the OAW Process

Conditions:

Vertical position
Refer to Exercise 2-5.

Performance:

The welder will demonstrate the correct procedure for welding a T-joint in the vertical position.

Criteria:

Welding technique, weld appearance, and weld strength as evaluated by the instructor.

Procedure:

1. Tack weld the two pieces to form a T-joint. Secure the welded piece in a positioner.
2. Start the weld at the bottom of the joint.
3. Hold the torch and filler rod at the same angle as in flat position.
4. Use a crescent motion with the torch to properly fill the joint. Direct the flame more toward the filler rod to control the puddle and filler metal deposited.

Exercise 2-16. Welding a Butt Joint in Overhead Position Using the OAW Process

Conditions:

Overhead position
Refer to Exercise 2-5.

Performance:

The welder will demonstrate the correct procedure for welding a butt joint in overhead position.

Criteria:

Welding technique, weld appearance, and weld strength as evaluated by the instructor.

Procedure:

1. Tack weld the two pieces to form a butt joint. Secure the welded piece in a positioner.
2. Use the same technique used in flat position, except use a circular motion with the filler rod. This motion helps distribute the deposited weld and prevents drops of metal from falling from the weld area.
3. Watch the flame closely. If the puddle begins to run, withdraw the torch.

Exercises 2-17 through 2-20. Welding Heavy Steel in the Flat Position Using the OAW Process

Conditions:

Flat position
$^1/_4$" × 2" × 5" for single V-butt joint and T-joint using the forehand technique
$^1/_2$" × 2" × 5" for double V-butt joint

Performance:

The welder will demonstrate the correct procedure for welding the following joints:

Exercise 2-17. Welding a Single V-Butt Joint Using the Forehand Technique
Exercise 2-18. Welding a Single V-Butt Joint Using the Backhand Technique
Exercise 2-19. Welding a Double V-Butt Joint Using the Forehand Technique
Exercise 2-20. Welding a T-Joint Using the Forehand Technique

Criteria:

Welding technique, weld appearance, and weld strength as evaluated by the instructor.

Procedure:

1. Bevel the edges of the plate as required.
2. Tack weld the two pieces, allowing a $^1/_{16}$" space between the two pieces.
3. Hold the torch at 60° from vertical rather than the 45° used on thinner metals.
4. Complete the welds using the specified technique.

Exercise 2-21. Welding Gray Cast Iron Using the OAW Process

Conditions:

Flat position
Gray cast iron

Performance:

The welder will demonstrate the correct procedure for welding gray cast iron.

Criteria:

Welding technique, weld appearance, and weld strength as evaluated by the instructor.

Procedure:

1. Prepare the edges to be welded.
2. Preheat the entire piece to a dull red.
3. Concentrate the flame at the starting point. When the metal begins to melt, move the flame from side to side to form a puddle in the entire V.
4. Heat the filler rod and dip it into the flux. Insert the fluxed end of the filler rod into the puddle. Do not dip the rod in and out of the puddle.
5. Move the puddle as the V is entirely filled. Repeat the operation until the joint is filled across the piece.

Exercise 2-22. Welding Aluminum In Flat Position Using the OAW Process

Conditions:

Flat position
Refer to Exercise 2-1.
$^3/_{16}$" × 2" × 5" aluminum

Performance:

The welder will demonstrate the correct procedure for welding the following joints:

Single V-butt joint
T-joint

Criteria:

Welding technique, weld appearance, and weld strength as evaluated by the instructor.

Procedure:

1. Prepare the edges of the pieces to be welded. Apply the recommended flux.
2. Pass the flame over the starting point until the flux melts.
3. Determine when welding should begin by scraping the filler rod over the surface of the metal.
4. Use a forehand technique, holding the torch 30° above horizontal.
5. The torch should be moved forward without any motion.
6. Dip the rod in and out of the weld puddle with a forward motion.
7. Maintain the same procedure across the weld joint.

SECTION ACTIVITIES 3

Chapters 12-23 Shielded Metal-Arc Welding — SMAW

1. Use arrows to show the flow of electrons in the following circuits:

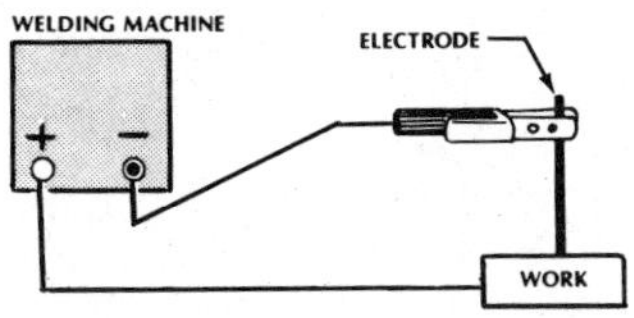

DC Straight Polarity

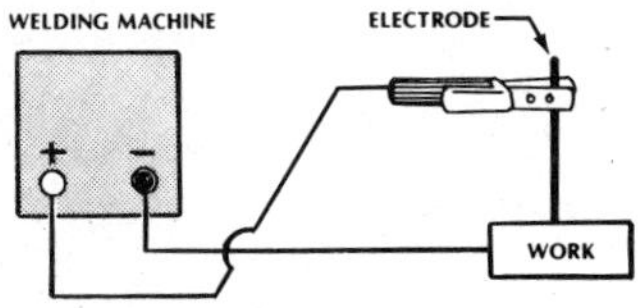

DC Reverse Polarity

2. List the type of current produced by the following welding machines:

Transformer ______________________________

Generator ______________________________

Rectifier ______________________________

3. Supply the characteristics pertaining to the electrodes used in shielded metal-arc welding.

AWS Classification	Weld Current	Position
EXX10		
EXX20		
EXX11		
EXX12		
EXX13		
EXX14		
EXX24		
EXX15		
EXX16		
EXX27		
EXX18		
EXX28		

Exercises 3-1 through 3-3. Striking the Arc in Shielded Metal-Arc Welding (SMAW)

Conditions:

Flat position
DC straight polarity

Exercise 3-1. E-6012 - 1/8″
Exercise 3-2. E-6013 - 1/8″
Exercise 3-3. E-7024 - 1/8″

1/4″ plate

Performance:

The welder will strike the arc using the scratching and/or tapping methods in the desired location.

Criteria:

The arc is struck in the correct location without excess spatter and without the electrode sticking to the plate.

Procedure:

1. Practice striking (starting) the arc using the methods shown.
2. Continue practicing until this operation can be performed quickly and easily.

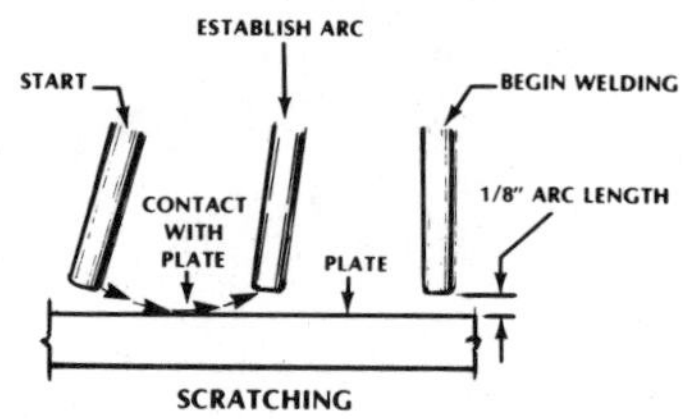

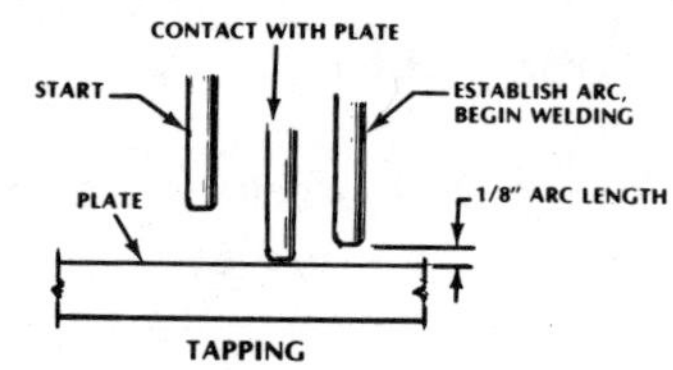

Exercises 3-4 through 3-6. Running Short Beads

Conditions:

Refer to Exercise 3-1.

Performance:

The welder will demonstrate the correct procedure for running short beads.

Criteria:

The beads deposited should be approximately 3/8" in width, and parallel to the length of the plate.

Procedure:

1. Mark the plate with soapstone as shown.
2. Deposit beads using a 90° work angle and a 15°-25° travel angle.

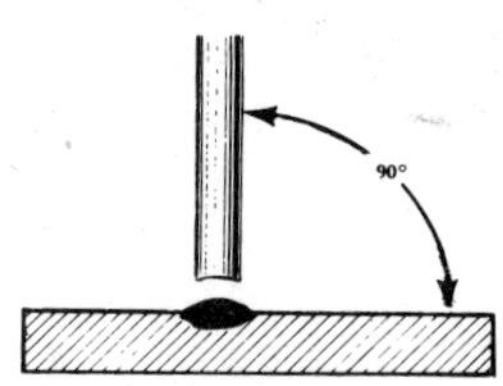

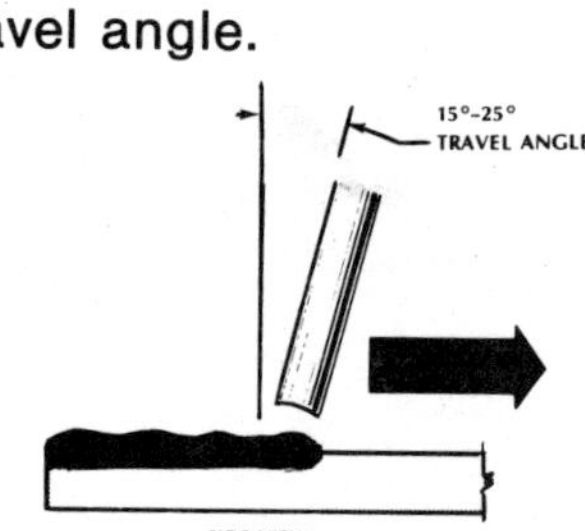

Exercises 3-7 through 3-11. Running Continuous Beads

Conditions:

Flat position
Exercise 3-7. E-6010 - 1/8" DC reverse polarity
Exercise 3-8. E-6011 - 1/8" DC reverse polarity
Exercise 3-9. E-6012 - 1/8" DC straight polarity
Exercise 3-10. E-6013 - 1/8" DC straight polarity
Exercise 3-11. E-7024 - 1/8" DC straight polarity
1/4" plate

Performance:

The welder will deposit continuous beads.

Criteria:

The beads should be consistent in width and parallel to the length of the plate.

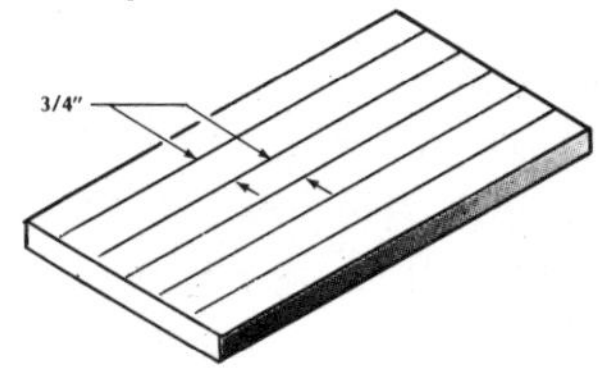

Procedure:

1. Use a soapstone to mark a series of lines $^3/_4$″ apart.
2. Deposit beads using the lines as a guide.
3. Remove the slag and examine the beads for consistent ripple formation.

Exercises 3-12 through 3-16. Moving the Electrode in Several Directions

Conditions:

Flat position

Exercise 3-12. E-6010 - $^1/_8$″ DC reverse polarity
Exercise 3-13. E-6011 - $^1/_8$″ DC reverse polarity
Exercise 3-14. E-6012 - $^1/_8$″ DC straight polarity
Exercise 3-15. E-6013 - $^1/_8$″ DC straight polarity
Exercise 3-16. E-7024 - $^1/_8$″ DC straight polarity

$^1/_4$″ plate

Performance:

The welder will deposit a continuous bead moving the electrode in different directions.

Criteria:

The bead deposited should show consistent bead formation resulting from correct arc length, work angle, travel angle, and speed of travel.

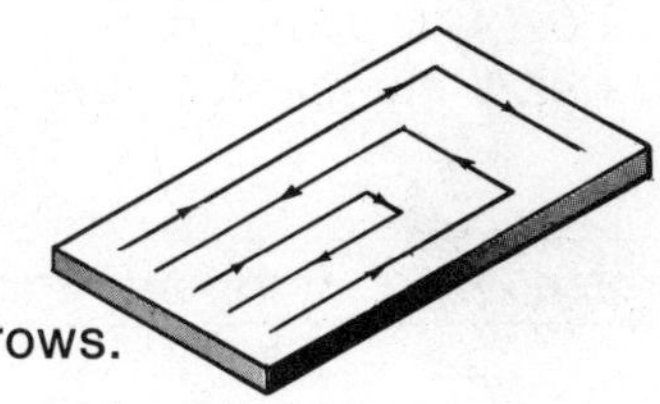

Procedure:

1. Draw lines on the plate as shown.
2. Deposit a continuous bead as shown by the direction of the arrows.

Exercises 3-17 through 3-21. Restarting the Arc

Conditions:

Flat position

Exercise 3-17. E-6010 - $^1/_8$″ DC reverse polarity
Exercise 3-18. E-6011 - $^1/_8$″ DC reverse polarity
Exercise 3-19. E-6012 - $^1/_8$″ DC straight polarity
Exercise 3-20. E-6013 - $^1/_8$″ DC straight polarity
Exercise 3-21. E-7024 - $^1/_8$″ DC straight polarity

$^1/_4$″ plate

Performance:

The welder will demonstrate the correct procedure for restarting an arc.

Criteria:

Consistent uniform beads are deposited with the preceding craters properly filled when the arc is restarted.

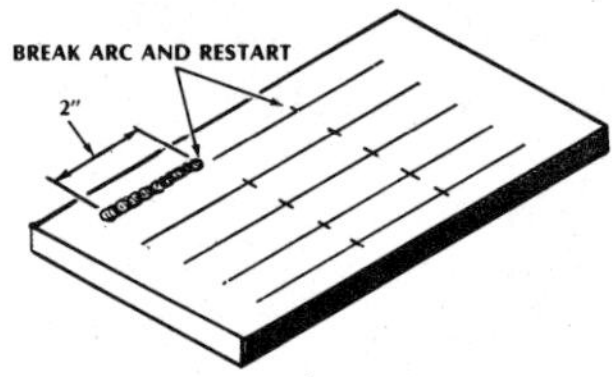

Procedure:

1. Draw a series of straight lines on the plate.
2. Divide the lines into 2″ sections.
3. Deposit a bead over the first 2″ section and break the arc.
4. Restart the arc and overlap the first 2″ section to completely fill the crater.

Exercises 3-22 through 3-24. Weaving the Electrode

Conditions:

Flat position

Exercise 3-22. E-6010 - 1/8″ DC reverse polarity
Exercise 3-23. E-6012 - 1/8″ DC straight polarity
Exercise 3-24. E-7024 - 1/8″ DC straight polarity

1/4″ plate

Performance:

The welder will demonstrate the correct technique for weaving the electrode.

Criteria:

The deposited bead is consistent in height, ripple shape, and penetration.

Procedure:

1. Lay out a series of lines parallel to the length of the plate at 1″ intervals.
2. Deposit straight beads over the lines.
3. Use a weaving motion to deposit a bead between the straight beads on the plate.
4. Use a crescent, figure 8, or rotary weaving motion as necessary with the electrodes used.

Exercise 3-25. Depositing a Metal Pad

Conditions:

Flat position

E-6010 - 1/8″ DC reverse polarity, layer 1
E-6011 - 1/8″ DC reverse polarity, layer 2
E-6012 - 1/8″ DC straight polarity, layer 3
E-6013 - 1/8″ DC straight polarity, layer 4
E-7024 - 1/8″ DC straight polarity, layer 5

3/8″ or thicker plate

Performance:

The welder will deposit successive overlapping beads to increase the thickness of the metal plate.

Criteria:

Each bead should overlap the previous bead in half, and completely penetrate the previous bead and the base metal with no voids in the weld metal applied.

Procedure:

1. Deposit a bead parallel to the length of the plate along the edge.
2. Remove the slag and deposit the next bead, which should penetrate half of the width of the previous bead and the base metal.
3. Repeat the operation until the entire piece is filled.
4. Using the designated electrode, deposit beads as described in Steps 1 and 2, parallel to the width of the piece.
5. Complete layers two through five, alternating the deposited beads by 90° for each layer.

Exercises 3-26 through 3-30. Welding a Butt Joint in Flat Position

Conditions:

Flat position

Exercise 3-26. E-6010 - $^{1}/_{8}$″ DC reverse polarity
Exercise 3-27. E-6011 - $^{1}/_{8}$″ DC reverse polarity
Exercise 3-28. E-6012 - $^{1}/_{8}$″ DC straight polarity
Exercise 3-29. E-6013 - $^{1}/_{8}$″ DC straight polarity
Exercise 3-30. E-7024 - $^{1}/_{8}$″ DC straight polarity

Two $^{3}/_{16}$″–$^{1}/_{4}$″ thick plates

Performance:

The welder will demonstrate the correct procedure for welding a butt joint in flat position.

Criteria:

Welding technique, weld appearance, and weld strength as evaluated by the instructor.

Procedure:

1. Butt the two pieces together and tack weld to form a butt joint.
2. Using a 90° work angle and a 15°–30° travel angle, weld the two butted pieces together.
3. Let cool and repeat the procedure on the reverse side.

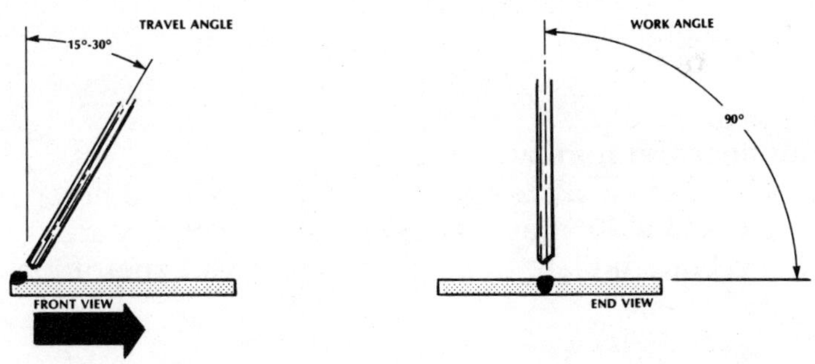

Exercises 3-31 through 3-35. Welding an Open Butt Joint in Flat Position

Conditions:

Flat position

Exercise 3-31. E-6010 - $^{1}/_{8}$″ DC reverse polarity
Exercise 3-32. E-6011 - $^{1}/_{8}$″ DC reverse polarity
Exercise 3-33. E-6012 - $^{1}/_{8}$″ DC straight polarity
Exercise 3-34. E-6013 - $^{1}/_{8}$″ DC straight polarity
Exercise 3-35. E-7024 - $^{1}/_{8}$″ DC straight polarity

Two $^{3}/_{16}$″–$^{1}/_{4}$″ plates

Performance:

The welder will demonstrate the correct procedure for welding an open butt joint in the flat position.

Criteria:

Welding technique, weld appearance, and weld strength as evaluated by the instructor.

Procedure:

1. Tack weld the two pieces to form a butt joint with a 1/8″ root opening. (Use the bare end of the electrode for consistent spacing.)
2. Refer to Exercises 3-26 through 3-30 for procedure.
3. Examine the butt weld. The weld bead should penetrate completely to the underside of the welded plates.

Exercises 3-36 through 3-40. Welding a Lap Joint in Flat Position

Conditions:

Flat position

Exercise 3-36. E-6010 - 1/8″ DC reverse polarity
Exercise 3-37. E-6011 - 1/8″ DC reverse polarity
Exercise 3-38. E-6012 - 1/8″ DC straight polarity
Exercise 3-39. E-6013 - 1/8″ DC straight polarity
Exercise 3-40. E-7024 - 1/8″ DC straight polarity

Two 3/16″–1/4″ plates

Performance:

The welder will demonstrate the correct procedure for welding a lap joint in flat position.

Criteria:

Refer to Exercises 3-26 through 3-30.

Procedure:

1. Tack weld the two pieces to form a lap joint.
2. Secure the tack welded piece in a positioner, or rest it against a firebrick to obtain flat position.
3. Use a 45° work angle and a 30° travel angle to deposit the bead.
4. Direct the heat to the bottom plate to prevent excessive penetration.

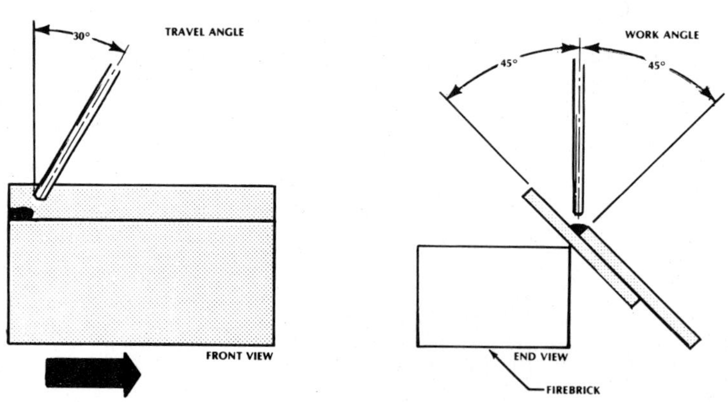

Exercises 3-41 through 3-45. Welding a Multiple-Pass Lap Joint in Flat Position

Conditions:

Flat position

Exercise 3-41. E-6010 - $^1/_8$" DC reverse polarity
Exercise 3-42. E-6011 - $^1/_8$" DC reverse polarity
Exercise 3-43. E-6012 - $^1/_8$" DC straight polarity
Exercise 3-44. E-6013 - $^1/_8$" DC straight polarity
Exercise 3-45. E-7024 - $^1/_8$" DC straight polarity

Two $^3/_{16}$"–$^1/_4$" plates

Performance:

The welder will demonstrate the correct procedure for welding a multiple-pass lap joint in the flat position.

Criteria:

Welding technique, weld appearance, and weld strength as evaluated by the instructor.

Procedure:

1. Refer to Exercises 3-36 through 3-40, and deposit a single pass on both sides of the lap joint.
2. Remove the slag.
3. Deposit the second pass using a weaving motion. Pause at the top and bottom of the weave as the weld is deposited.

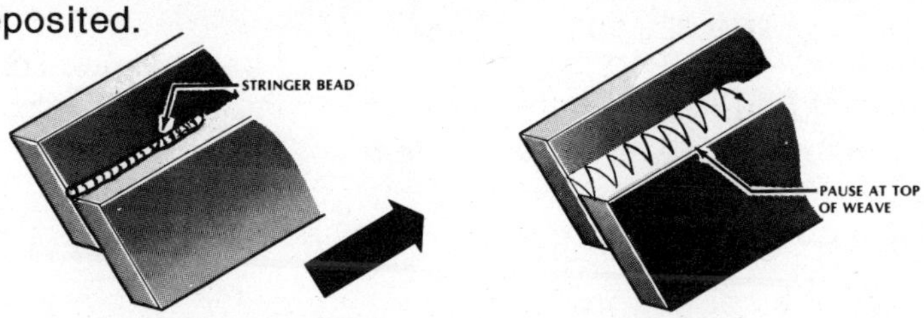

Exercises 3-46 through 3-50. Welding a T-Joint in Flat Position

Conditions:

Refer to Exercises 3-31 through 3-35.

Performance:

The welder will demonstrate the correct procedure for welding a T-joint in flat position.

Criteria:

Welding technique, weld appearance, and weld strength as evaluated by the instructor.

Procedure:

1. Tack weld the two pieces to form a T-joint.
2. Secure the tack welded piece in a positioner or rest it against a firebrick to obtain flat position.
3. Use a 45° work angle and a 30° travel angle to deposit the bead.
4. Advance the electrode as the proper fillet contour is formed.

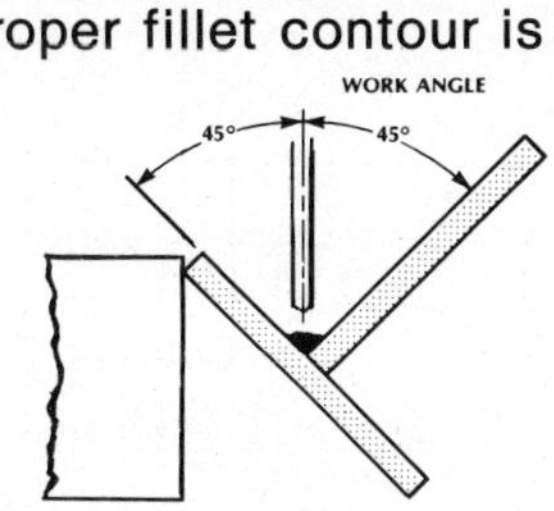

Exercises 3-51 through 3-55. Welding a Multiple-Pass T-Joint in Flat Position

Conditions:

Refer to Exercises 3-31 through 3-35.

Performance:

The welder will demonstrate the correct procedure for welding a multiple-pass T-joint.

Criteria:

Welding technique, weld appearance, and weld strength as evaluated by the instructor.

Procedure:

1. Refer to Exercises 3-46 through 3-50, and deposit a single pass on both sides of the T-joint.
2. Remove the slag.

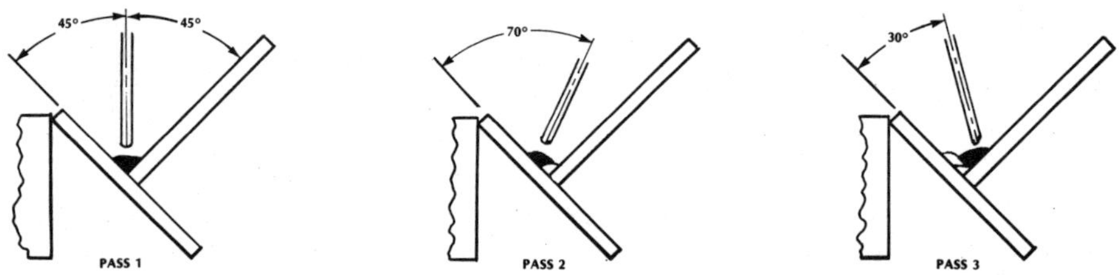

3. Deposit the second pass using a 70° work angle and a 30° travel angle. The bead should penetrate half into the base metal and half into the root pass. Remove the slag.
4. Deposit the third pass using a 30° work angle and a 30° travel angle. Half of the bead should penetrate into the base metal and half into the second pass.
5. If more weld metal is required, additional passes may be deposited as shown.

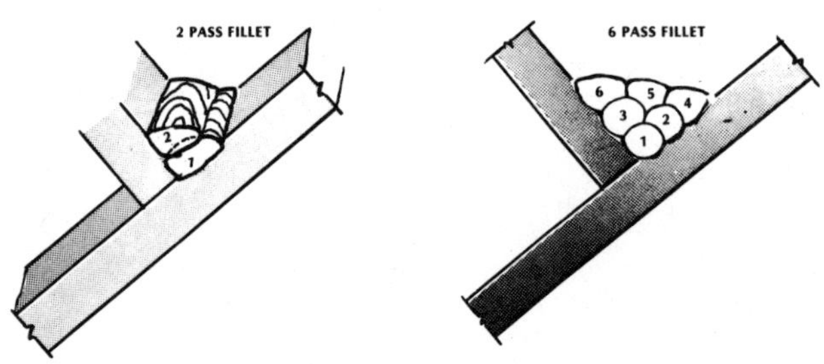

Exercises 3-56 through 3-60. Welding a Multiple-Pass T-Joint in Flat Position With a Cover Pass

Conditions:

Refer to Exercises 3-31 through 3-35.

Performance:

The welder will demonstrate the correct procedure for welding a multiple-pass T-joint in flat position with a cover pass.

Criteria:

Welding technique, weld appearance, and weld strength as evaluated by the instructor.

Procedure:

1. Refer to Exercises 3-16 through 3-35 and deposit a single pass on both sides of the T-joint.
2. Remove the slag.
3. Deposit the second pass using a 70° work angle and a 30° travel angle. Half of the bead should penetrate into the base metal and half into the root pass. Remove the slag.
4. Deposit the third pass using a 30° work angle and a 30° travel angle. Half of the bead should penetrate into the base metal and half into the second pass. Remove the slag.
5. Use a weaving motion to deposit the cover pass on the joint. Pause slightly at the toes of the weld to prevent undercutting.

Exercises 3-61 through 3-65. Welding a Corner Joint in Flat Position

Conditions:

Refer to Exercises 3-31 through 3-35.

Performance:

The welder will demonstrate the correct procedure for welding a corner joint in flat position.

Criteria:

Welding technique, weld appearance, and weld strength as evaluated by the instructor.

Procedure:

1. Tack weld the pieces to form an open corner joint.
2. Secure the tack welded piece in a positioner or rest it on the bench to obtain flat position.
3. Use a 45° work angle and a 30° travel angle to deposit a fillet weld across the joint.
4. If additional weld metal is required, deposit a second pass using a weaving motion.

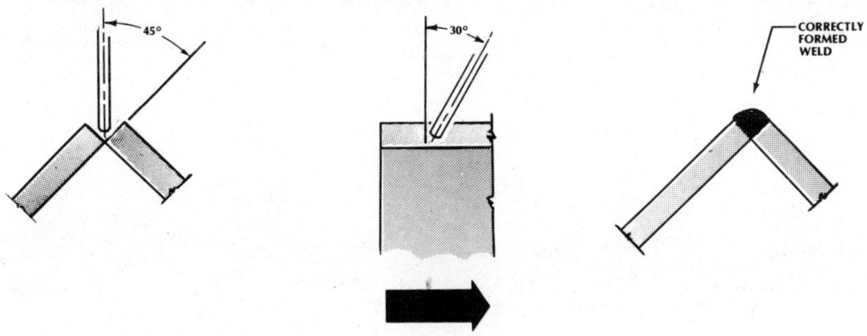

Exercises 3-66 and 3-67. Welding Round Stock in Flat Position

Conditions:

Flat position

Exercise 3-66. E-6010 - $^1/_8$″ DC reverse polarity
Exercise 3-67. E-6013 - $^1/_8$″ DC straight polarity

Round stock

Performance:

The welder will demonstrate the correct procedure for welding round stock in the flat position.

Criteria:

Welding technique, weld appearance, and weld strength as evaluated by the instructor.

Procedure:

1. Bevel both sides of the round stock, leaving a root face in the center. Both sides should have the same groove angle.
2. Place the pieces in a vise or a section of angle iron to maintain the required welding position.
3. Tack weld both sides.
4. Deposit a root pass on one side, then the other side.
5. Use a weaving motion to fill the groove completely.

Exercises 3-68 through 3-70. Depositing Beads in Horizontal Position

Conditions:

Horizontal position

Exercise 3-68. E-6010 - 1/8″ DC reverse polarity
Exercise 3-69. E-6012 - 1/8″ DC straight polarity
Exercise 3-70. E-7018 - 1/8″ DC reverse polarity

1/4″ × 4″ × 6″ plate

Performance:

The welder will deposit stringer beads in the horizontal position.

Criteria:

The beads should be 5/16″ to 3/8″ wide and parallel to the edge of the plate.

Procedure:

1. Draw a series of lines 1/2″ apart and parallel to the length of the plate.

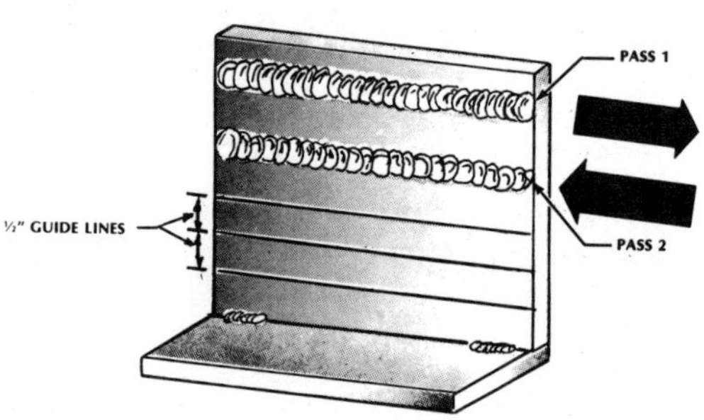

2. Tack weld to a flat plate or secure the piece in a positioner.
3. Using the guidelines drawn, deposit straight, consistent beads with a 5°–10° work angle and a 20° travel angle.

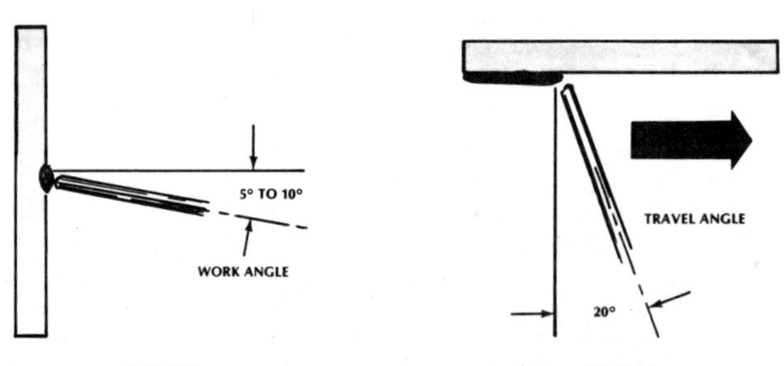

Exercises 3-71 through 3-73. Depositing a Metal Pad in Horizontal Position

Conditions:

Horizontal Position

Exercise 3-71. E-6010 - 1/8″ DC reverse polarity
Exercise 3-72. E-6013 - 1/8″ DC reverse polarity
Exercise 3-73. E-7018 - 1/8″ DC reverse polarity

1/4″-3/8″ × 3″ × 4″ plate

Performance:

The welder will deposit successive overlapping beads to increase the thickness of the metal plate.

Criteria:

Each bead should overlap the previous bead in half and completely penetrate the previous bead and the base metal with no voids in the weld metal applied.

Procedure:

1. Tack weld to a flat plate or secure the piece in a positioner.
2. Deposit a bead along the edge of the plate and parallel to its length.
3. Remove the slag and deposit the next bead, which should penetrate half the width of the previous bead and the base metal.
4. Repeat the operation until the entire piece is filled.

Exercises 3-74. Welding a Lap Joint in Horizontal Position

Conditions:

Horizontal position
Adjust to correct amperage
E-6010 - 1/8″ DC reverse polarity
Two 1/4″ × 1 1/2″ × 6″

Performance:

The welder will demonstrate the correct procedure for welding a lap joint in horizontal position.

Criteria:

Welding technique, weld appearance, and weld strength as evaluated by the instructor.

Procedure:

1. Tack weld the two pieces to form a lap joint.
2. Secure the tack welded piece in a positioner.
3. Use a 45° work angle and a 15°–30° travel angle.

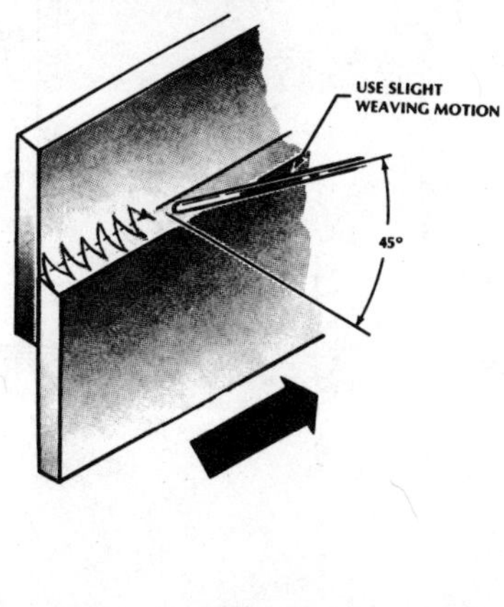

4. Use a slight weaving motion, pausing at the toes to avoid undercutting.
5. Remove the slag and deposit a second pass with a weaving motion.
6. Repeat the exercise on the other side.

Exercise 3-75. Welding a T-Joint in Horizontal Position

Conditions:

Horizontal position
E-6010 - $^1/_8$″ DC reverse polarity
Two $^1/_4$″ × 2″ × 4″ plates

Performance:

The welder will demonstrate the correct procedure for welding a T-joint in horizontal position.

Criteria:

Welding technique, weld appearance, and weld strength as evaluated by the instructor.

Procedure:

1. Tack weld the two pieces to form a T-joint.
2. Use a 45° work angle and a 15°–30° travel angle.
3. Deposit a single bead in the root of the weld with no weaving motion.
4. Repeat the exercise on the other side.

Exercise 3-76. Welding a Multiple-Pass T-Joint in Horizontal Position

Conditions:

Refer to Exercise 3-75.

Performance:

The welder will demonstrate the correct procedure for welding a multiple-pass T-joint in the horizontal position.

Criteria:

Welding technique, weld appearance, and weld strength as evaluated by the instructor.

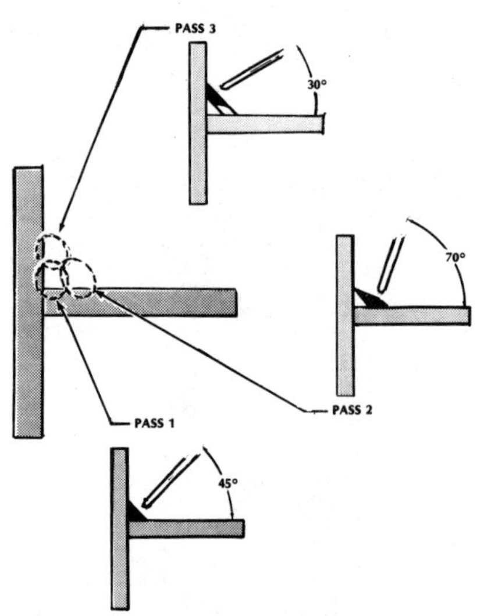

Procedure:

1. Tack weld the two pieces to form a T-joint.
2. Use a 45° work angle and a 15°–30° travel angle to deposit the first pass.
3. Remove the slag. Use a 70° work angle and a 15°–30° travel angle to deposit the second pass. Half of the bead should penetrate into the root pass and half into the base metal.
4. Remove the slag. Use a 30° work angle and a 15°–30° travel angle to deposit the third pass. Half of the bead should penetrate into the second pass and half into the base metal.
5. Repeat the exercise on the other side.

Exercises 3-77 through 3-81. Depositing Beads in the Vertical Position (Vertical Down)

Conditions:

Vertical position

Exercise 3-77. E-6010 - 1/8″ DC reverse polarity
Exercise 3-78. E-6011 - 1/8″ DC reverse polarity
Exercise 3-79. E-6012 - 1/8″ DC reverse polarity
Exercise 3-80. E-6013 - 1/8″ DC reverse polarity
Exercise 3-81. E-7018 - 1/8″ DC reverse polarity

1/4″–3/8″ plate

Performance:

The welder will deposit beads in the vertical position using the vertical down technique.

Criteria:

The deposited bead should be consistent in height, ripple shape, and penetration.

Procedure:

1. Draw a series of guidelines parallel to the length of the plate.
2. Start at the top using a 90° work angle and a 30° travel angle to deposit straight beads.
3. Maintain a short arc length.

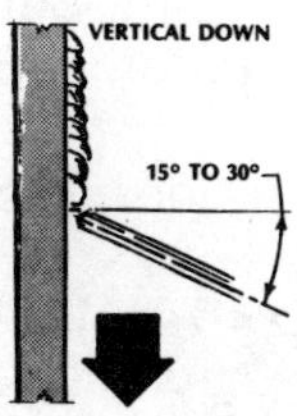

VERTICAL DOWN

Exercises 3-82 through 3-84. Depositing Beads in the Vertical Position (Vertical Up)

Conditions:

Vertical position

Exercise 3-82. E-6010 - 1/8″ DC reverse polarity
Exercise 3-83. E-6011 - 1/8″ DC reverse polarity
Exercise 3-84. E-7018 - 1/8″ DC reverse polarity

1/4″–3/8″ plate

Performance:

The welder will deposit beads in the vertical position using the vertical up technique.

Criteria:

The deposited bead should be consistent in height, ripple shape, and penetration.

Procedure:

1. Draw a series of guidelines parallel to the length of the plate.
2. Use a 90° work angle. To control the heat and bead formation, use a whipping motion with the E-6010 and E-6011 electrodes. Use a slight side-to-side weave with the E-7018 electrode.
3. Maintain a short arc length.

Exercises 3-85 and 3-86. Welding a Lap Joint in Vertical Position (Vertical Up)

Conditions:

Vertical position

Exercise 3-85. E-6010 - 1/8" DC reverse polarity
Exercise 3-86. E-7018 - 1/8" DC reverse polarity

Two 1/4" × 2" × 4" plates

Performance:

The welder will demonstrate the correct procedure for welding a two-pass lap joint in vertical position (vertical up).

Criteria:

Welding technique, weld appearance, and weld strength as evaluated by the instructor.

Procedure:

1. Tack weld the two pieces to form a lap joint. Secure the tack welded piece in a positioner.
2. Use a 45° work angle and a whipping motion (E-6010), or a slight weaving motion (E-7018) to deposit the root pass.
3. Remove the slag and deposit the second pass using a weaving motion. Pause at the toes of the weld to eliminate undercutting.

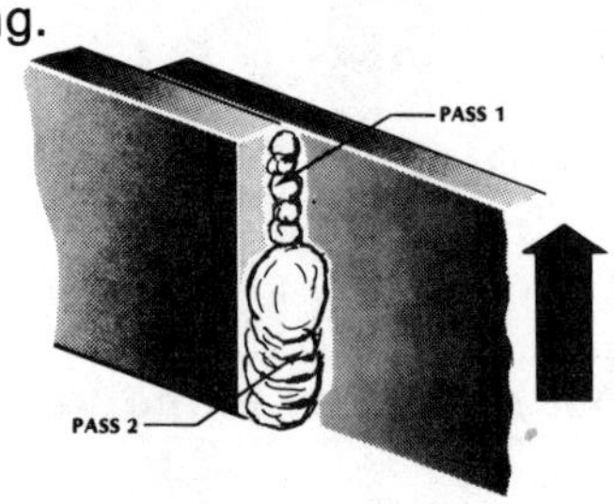

Exercises 3-87 and 3-88. Welding a Single V-Butt Joint in Vertical Position (Vertical Up)

Conditions:

Vertical position

Exercise 3-87. E-6010 - 1/8" DC reverse polarity
Exercise 3-88. E-7018 - 1/8" DC reverse polarity

Two 1/4" × 4" × 6" plates

Performance:

The welder will demonstrate the correct procedure for welding a single V-butt joint in the vertical position (vertical up).

Criteria:

Welding technique, weld appearance, and weld strength as evaluated by the instructor.

Procedure:

1. Bevel the edges to form a 60° included angle. Tack weld the two pieces to form a butt joint. Allow a 1/16″ root opening.
2. Secure the tack welded piece in a positioner.
3. Deposit the root pass and penetrate completely to the opposite side. Remove the slag. Use a whipping motion (E-6010) or a drag motion (E-7018).
4. Increase the amperage slightly and deposit the second pass, completely penetrating the root pass and the base metal. Remove the slag.
5. Reduce the amperage and deposit the third pass (cover pass) with a weaving motion.

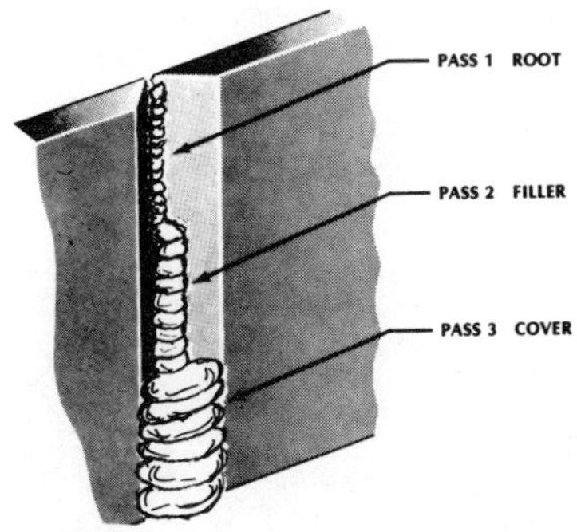

Exercises 3-89 and 3-90. Welding a Multiple-Pass T-Joint in Vertical Position (Vertical Up)

Conditions:

Vertical position

Exercise 3-89. E-6010 - 1/8″ DC reverse polarity
Exercise 3-90. E-7018 - 1/8″ DC reverse polarity

Two 1/4″ × 4″ × 6″ plates

Performance:

The welder will demonstrate the correct procedure for welding a multiple-pass T-joint in vertical position (vertical up).

Criteria:

Welding technique, weld appearance, and weld strength as evaluated by the instructor.

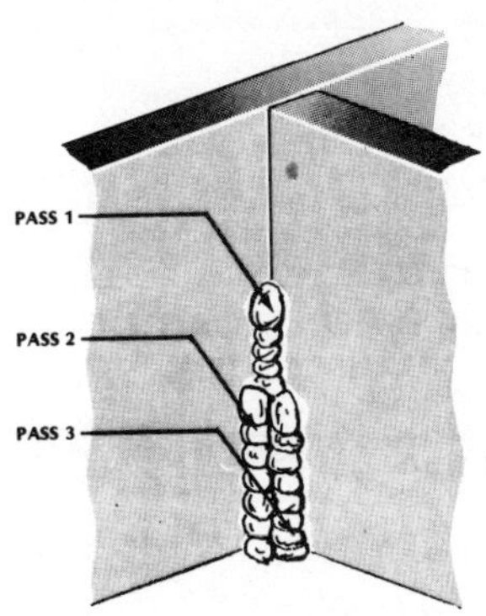

Procedure:

1. Tack weld the two pieces to form a T-joint. Secure the tack welded piece in a positioner.
2. Deposit the root pass using a 45° work angle. Remove the slag.
3. Deposit the second pass using a 70° work angle. Remove the slag.
4. Deposit the third pass using a 30° work angle.

Exercises 3-91 through 3-93. Depositing Beads in Overhead Position

Conditions:

Overhead position

Exercise 3-91. E-6010 - 1/8″ DC reverse polarity
Exercise 3-92. E-6011 - 1/8″ DC reverse polarity
Exercise 3-93. E-7018 - 1/8″ DC reverse polarity

1/4″–3/8″ plate

Performance:

The welder will deposit beads in the overhead position.

Criteria:

The deposited bead should be consistent in height, ripple shape, and penetration.

Procedure:

1. Draw a series of guidelines ½″ apart, parallel to the length of the plate.

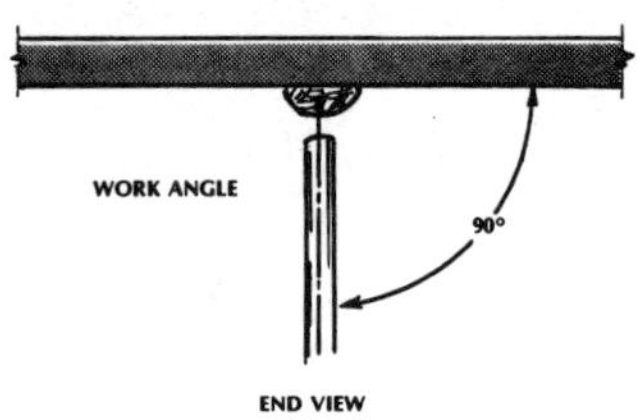

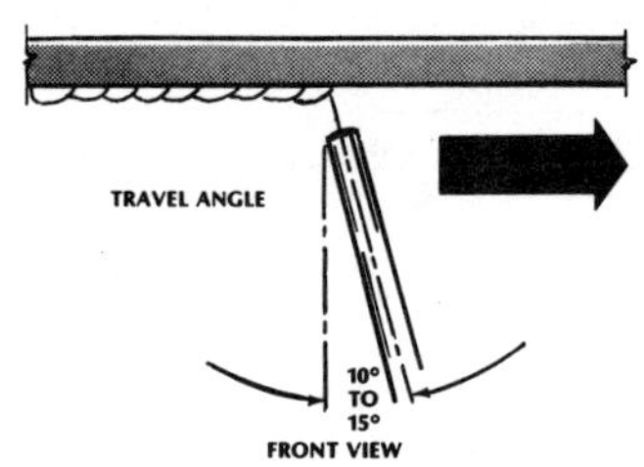

2. Secure the plate in a positioner to obtain overhead position.
3. Reduce the amperage as recommended, and deposit straight beads over the guidelines. Reverse the direction.
4. After depositing beads across the plate, fill in the space between the beads using a weaving motion.

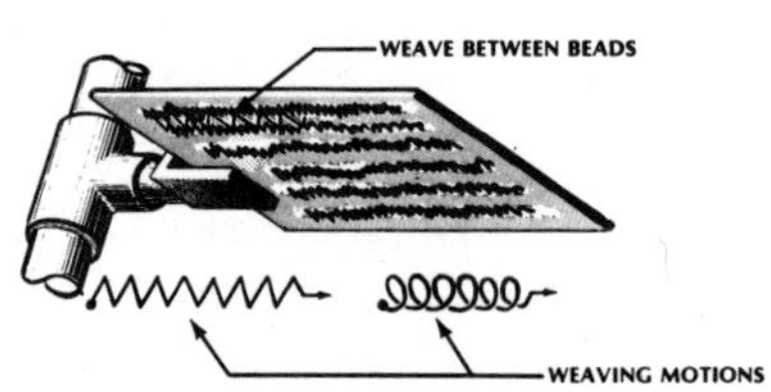

Exercises 3-94 and 3-95. Welding a Multiple-Pass Lap Joint in Overhead Position

Conditions:

Overhead position

Exercise 3-94. E-6010 - 1/8″ DC reverse polarity
Exercise 3-95. E-7018 - 1/8″ DC reverse polarity

Two 1/4″ × 2″ × 4″ plates

Performance:

The welder will demonstrate the correct procedure for welding a multiple-pass lap joint in the overhead position.

Criteria:

Welding technique, weld appearance, and weld strength as evaluated by the instructor.

Procedure:

1. Tack weld the pieces to form a lap joint. Secure the tack welded piece in a positioner. Reduce the amperage as recommended.
2. Use a 45° work angle and a 15° travel angle to deposit the first pass. Remove the slag.
3. Deposit the second pass, penetrating the root pass and the base metal. Adjust the work angle as necessary.
4. Deposit the third pass, penetrating the second pass and the base metal. Adjust the work angle as necessary.

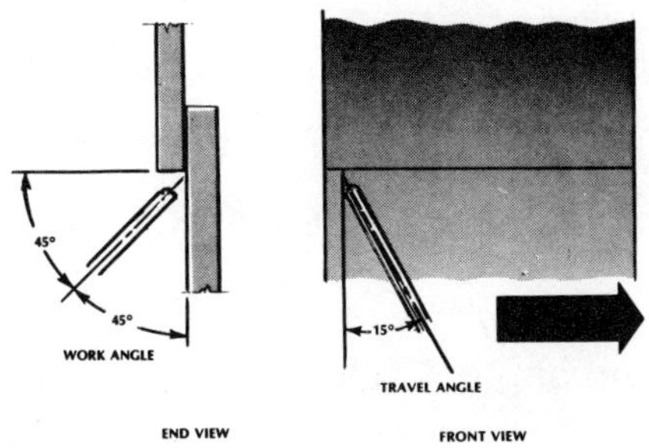

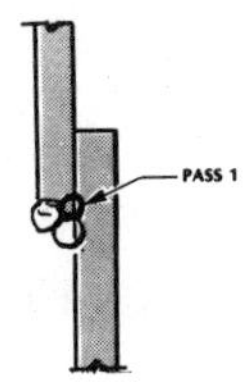

Exercises 3-96 and 3-97. Welding a Multiple-Pass T-Joint in Overhead Position

Conditions:

Overhead position

Exercise 3-96. E-6010 - 1/8″ DC reverse polarity
Exercise 3-97. E-7018 - 1/8″ DC reverse polarity

Two 1/4″ × 2″ × 4″ plates

Performance:

The welder will demonstrate the correct procedure for welding a multiple pass T-joint in the overhead position.

Criteria:

Welding technique, weld appearance, and weld strength as evaluated by the instructor.

Procedure:

1. Tack weld the pieces to form a T-joint. Secure the tack welded piece in a positioner. Reduce the amperage as recommended.

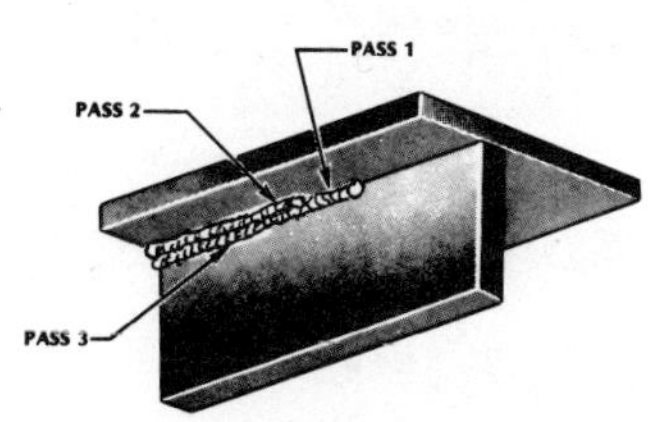

2. Use a 45° work angle and a 15° travel angle to deposit the first pass. Remove the slag.
3. Deposit the second pass, penetrating the base metal. Adjust the work angle as necessary. Remove the slag.
4. Deposit the third pass, penetrating the second pass and the base metal.

Exercises 3-98 and 3-99. Welding a Single V-Butt Joint in Overhead Position

Conditions:

Overhead position

Exercise 3-98. E-6010 - 1/8" DC reverse polarity
Exercise 3-99. E-6011 - 1/8" DC reverse polarity

Two 1/4" × 2" × 4" plates

Performance:

The welder will demonstrate the correct procedure for welding a single V-butt joint in overhead position.

Criteria:

Welding technique, weld appearance, and weld strength as evaluated by the instructor.

Procedure:

1. Bevel the edges to form a 60° included angle. Tack weld the two pieces to form a butt joint. Allow a 1/16" root opening.
2. Secure the tack welded piece in positioner.
3. Deposit the root pass and obtain complete penetration to the opposite side. Remove the slag.
4. Deposit two additional passes to cover the groove faces of the joint.

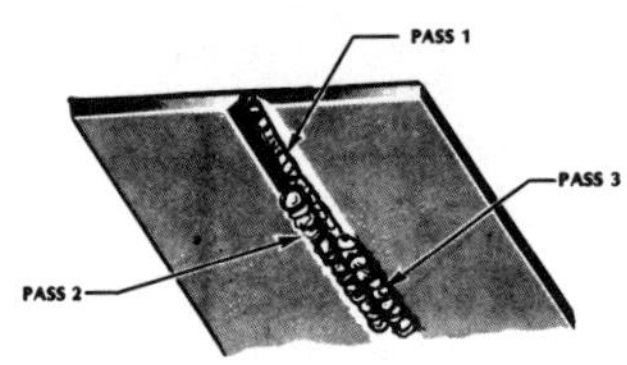

SECTION ACTIVITIES 4

Chapters 24–26 Gas Shielded-Arc Welding

Exercise 4-1. Running Beads on Mild Steel in Flat Position Using the GTAW Process Without Filler Rod

Conditions:

Flat position
DC straight polarity
50–60 amps
$\frac{3}{32}''$, 1% thoriated tungsten electrode (tip pointed) $\frac{1}{8}''$–$\frac{3}{16}''$ electrode extension
Argon shielding gas 20 cfm with 15-second post purge
Weld current remote off, contactor control on
16-gauge mild steel 4″ × 6″

Performance:

The welder will deposit continuous beads without a filler rod.

Criteria:

The beads should be approximately $\frac{1}{8}''$ wide and parallel to the length of the plate.

Procedure:

1. To start the arc, activate the current flow by pushing the foot pedal or turning on the switch at the torch.
2. Position the torch at a 45° angle with the electrode $\frac{1}{8}''$ from the plate.
3. When the arc starts, raise the torch to a 90° work angle and a 20° push angle.
4. Maintain a close arc while running beads approximately $\frac{3}{8}''$ apart.

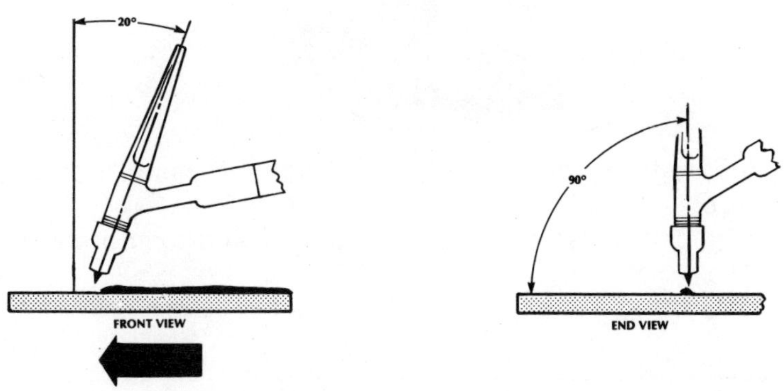

Exercise 4-2. Running Beads on Mild Steel in Flat Position Using the GTAW Process with Filler Rod

Conditions:

Flat position
Refer to Exercise 4-1.
$\frac{1}{16}''$ recommended filler rod

Performance:

The welder will deposit continuous beads using a filler rod.

Criteria:

The beads should be approximately 3/16″ wide, convex in shape, and parallel to the length of the plate.

Procedure:

1. Refer to Exercise 4-1, Steps 1–4.
2. While maintaining the arc, hold the filler rod at a 20° angle. Dip the filler rod into the leading edge of the puddle using an in-and-out motion.
3. Use a small rotary motion to form a bead approximately 3/16″ wide.
4. Run a series of straight consistent beads 3/8″ apart across the plate.

Exercise 4-3. Welding a Butt Joint on Mild Steel in Flat Position Using the GTAW Process

Conditions:

Flat position
Refer to Exercise 4-1.
Two pieces of 16-gauge 1 1/2″ × 6″
1/16″ recommended filler rod

Performance:

The welder will demonstrate the correct procedure for welding a butt joint in flat position.

Criteria:

Welding technique, weld appearance, and weld strength as evaluated by the instructor.

Procedure:

1. Tack weld the two pieces of steel to form a butt joint with no root opening.

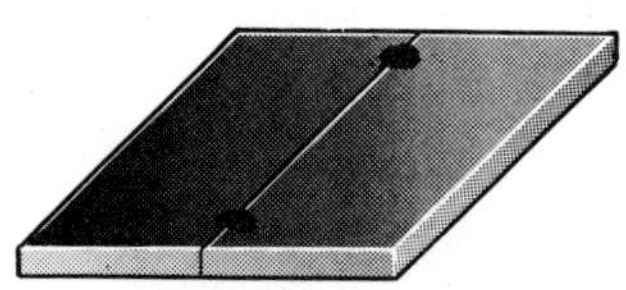

2. Use the same procedure for running beads with filler rod using the joint as the center of the weld across the piece.
3. Penetration should be complete with a bead width of approximately 3/16″.

Exercise 4-4. Welding a Lap Joint on Mild Steel in Horizontal Position Using the GTAW Process

Conditions:

Horizontal position
Refer to Exercise 4-1.
Two pieces of 16-gauge 1 1/2″ × 6″
1/16″ recommended filler rod

Performance:

The welder will demonstrate the correct procedure for welding a lap joint in horizontal position.

Criteria:

Welding technique, weld appearance, and weld strength as evaluated by the instructor.

Procedure:

1. Tack weld the two pieces of steel to form a lap joint.
2. Hold the torch at a 80°–85° work angle and a 20° push angle.

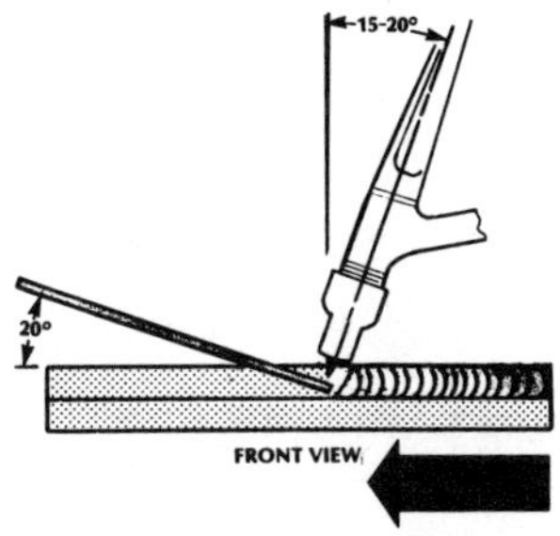

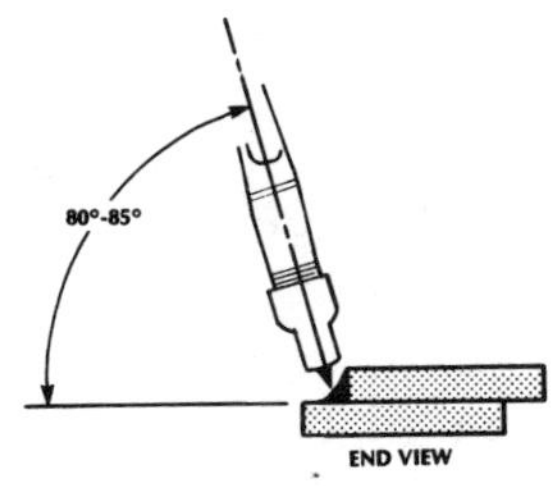

3. Add filler metal using an in-and-out motion while directing the arc to the bottom plate.
4. Carefully maintain a consistent bead across the piece.

Exercise 4-5. Welding a T-Joint on Mild Steel in Horizontal Position Using the GTAW Process

---

Conditions:

Horizontal position
Refer to Exercise 4-1.
Two pieces of 16-gauge 1 1/2″ × 6″
1/16″ filler rod

Performance:

The welder will demonstrate the correct procedure for welding a T-joint in horizontal position.

Criteria:

Welding technique, weld appearance, and weld strength as evaluated by the instructor.

Procedure:

1. Tack weld the two pieces to form a T-joint.
2. Hold the torch at a 45° work angle and a 15° push angle.

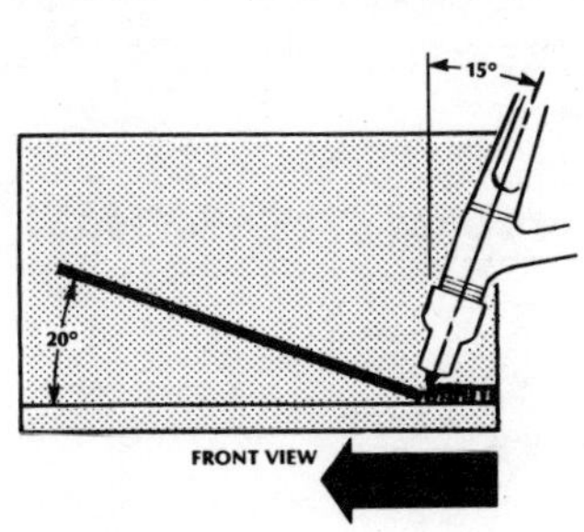

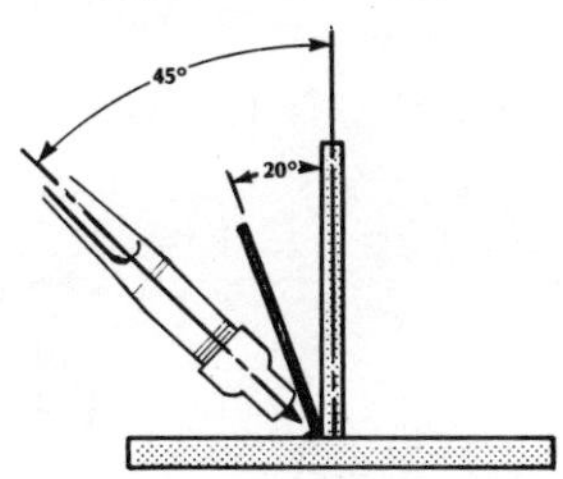

3. Add filler metal by holding the filler rod at a 20° angle from the bottom plate, and a 20° angle from the top plate.
4. Weave the torch slightly while adding filler metal with an in-and-out motion.
5. Carefully maintain a consistent bead across the piece.

Exercise 4-6. Welding a Butt Joint on Mild Steel in Horizontal Position Using the GTAW Process

Conditions:

Horizontal position
Refer to Exercise 4-1.
Two pieces of 16-gauge $1^{1}/_{2}$″ × 6″
$^{1}/_{16}$″ filler rod

Performance:

The welder will demonstrate the correct procedure for welding a butt joint in horizontal position.

Criteria:

Welding technique, weld appearance, and weld strength as evaluated by the instructor.

Procedure:

1. Tack weld the two pieces to form a butt joint. Secure the tack welded piece in a positioner.
2. Hold the torch at a 15° work angle and a 15° push angle.
3. Position the filler rod at a 20° angle in line with the weld bead.
4. Weave the torch slightly while adding filler rod with an in-and-out motion on the top half of the leading edge of the puddle.
5. Carefully maintain a consistent bead across the piece.

Exercise 4-7. Welding a Butt Joint on Mild Steel in Vertical Position Using the GTAW Process

Conditions:

Vertical position
Refer to Exercise 4-1.
Two pieces of 12-gauge $1^{1}/_{2}$″ × 6″
$^{1}/_{16}$″ filler rod

Performance:

The welder will demonstrate the correct procedure for welding a butt joint in the vertical position.

Criteria:

Welding technique, weld appearance, and weld strength as evaluated by the instructor.

Procedure:

1. Tack weld the two pieces to form a butt joint. Secure the tack welded piece in a positioner.

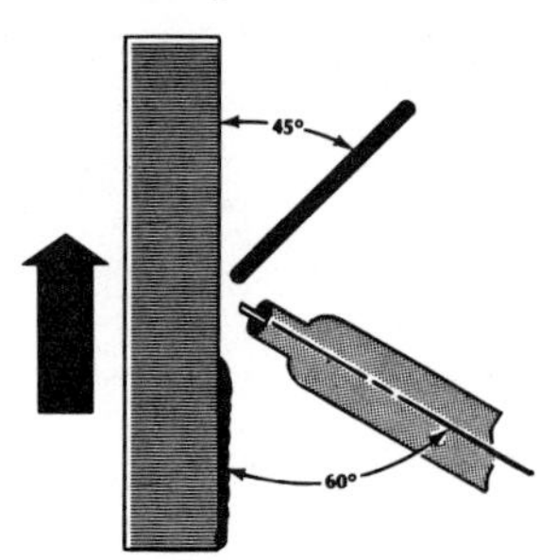

2. Hold the torch at a 90° work angle and a 60° push angle.
3. Position the filler rod at a 45° angle in line with the weld bead.
4. Use a slight weaving motion while adding filler metal with an in-and-out motion.
5. Carefully maintain a consistent bead across the piece.

Exercise 4-8. Welding a T-Joint on Mild Steel in Vertical Position Using the GTAW Process

Conditions:

Vertical position
Refer to Exercise 4-1.
Two pieces of 16-gauge $1^1/_2'' \times 6''$
$^1/_{16}''$ filler rod

Performance:

The welder will demonstrate the correct procedure for welding a T-joint in vertical position.

Criteria:

Welding technique, weld appearance, and weld strength as evaluated by the instructor.

Procedure:

1. Tack weld the two pieces to form a T-joint. Secure the tack welded piece in a positioner.
2. Hold the torch at a 45° work angle and a 20° push angle.

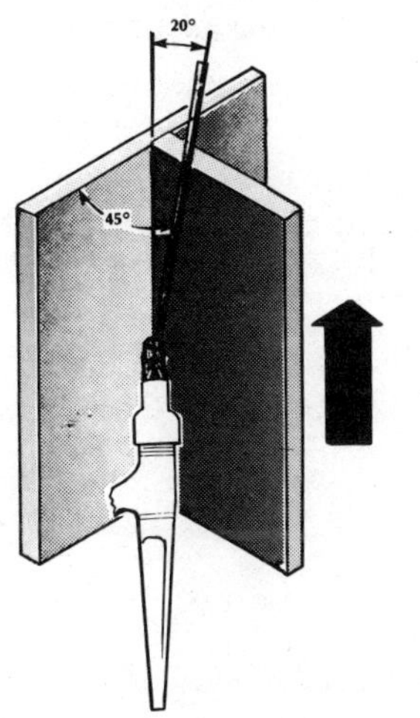

3. Position the filler rod at a 20° angle centered between the parts of the weld.
4. Weave the torch slightly while adding filler metal with an in-and-out motion.
5. Carefully maintain a consistent bead across the piece.

Exercise 4-9. Welding a Butt Joint on Mild Steel in Overhead Position Using the GTAW Process

Conditions:

Overhead position
Refer to Exercise 4-1.
Two pieces of 16-gauge $1^1/_2'' \times 6''$
$^1/_{16}''$ filler rod

Performance:

The welder will demonstrate the correct procedure for welding a butt joint in the overhead position.

Criteria:

Welding technique, weld appearance, and weld strength as evaluated by the instructor.

Procedure:

1. Tack weld the two pieces to form a butt joint. Secure the tack welded piece in a positioner.
2. Reduce the amperage 5%–10% from the amount used for flat position.
3. Use the same procedure that is used for the flat position. Refer to Exercise 4-3.

Exercise 4-10. Welding a Lap Joint on Mild Steel in Overhead Position Using the GTAW Process

Conditions:

Overhead position
Refer to Exercise 4-1.
Two pieces of 16-gauge $1^1/_2'' \times 6''$
$^1/_{16}''$ filler rod

Performance:

The welder will demonstrate the correct procedure for welding a lap joint in the overhead position.

Criteria:

Welding technique, weld appearance, and weld strength as evaluated by the instructor.

Procedure:

1. Tack weld the two pieces to form a lap joint. Secure the welded piece in a positioner.
2. Reduce the amperage 5%–10% from the amount used for flat position.
3. Refer to Exercise 4-4 for procedure.

Exercise 4-11. Welding a T-Joint on Mild Steel in Overhead Position Using the GTAW Process

Conditions:

Overhead position
Refer to Exercise 4-1.
Two pieces of 16-gauge $1^1/_2'' \times 6''$ mild steel
$^1/_{16}''$ filler rod

Performance:

The welder will demonstrate the correct procedure for welding a T-joint in the overhead position.

Criteria:

Welding technique, weld appearance, and weld strength as evaluated by the instructor.

Procedure:

1. Tack weld the two pieces to form a T-joint. Secure the tack welded piece in a positioner.
2. Reduce the amperage 5%–10% from amount used for flat position.
3. Refer to Exercise 4-5 for procedure.

Exercise 4-12. Running Beads on Aluminum in Flat Position Using the GTAW Process

Conditions:

Flat position
AC high frequency should be set for continuous
Weld current remote contactor control on
140–150 amps

$^{3}/_{32}$" pure tungsten electrode (tip spherical), $^{1}/_{8}$"–$^{3}/_{16}$"electrode extension
Argon shielding gas 20 cfm with 15-second post purge extension
$^{1}/_{8}$" × 4" × 6" aluminum
$^{1}/_{8}$" recommended filler rod

Performance:

The welder will demonstrate the correct procedure for running beads on aluminum using the GTAW process.

Criteria:

Welding technique, weld appearance, and weld strength as evaluated by the instructor.

Procedure:

1. Clean the aluminum pieces with a stainless steel wire brush.
2. To form the spherical end on the electrode, switch the welding machine to DCRP and strike the arc on a copper plate while holding the welding torch at a 90° angle.
3. Start the arc by positioning the electrode $^{1}/_{8}$" away from the work and activating the current flow by pushing the foot pedal control or the switch at the torch. Do not touch the electrode to the work.
4. Hold the torch at a 90° work angle and a 20° push angle.
5. Melt the base metal and add filler metal to form a bead $^{1}/_{4}$" wide. Add filler rod to the leading edge of the puddle using an in-and-out motion. Weave the torch slightly to distribute the heat to form the bead.
6. To fill the crater at the end of the weld, reduce the amperage and continue to add filler rod.
7. Run a series of beads on the piece approximately $^{3}/_{8}$" apart.

Exercise 4-13. Welding a Butt Joint on Aluminum in Flat Position Using the GTAW Process

Conditions:

Flat position
Refer to Exercise 4-12.
Two pieces of $^{1}/_{8}$" × 1$^{1}/_{2}$" × 6" aluminum

Performance:

The welder will demonstrate the correct procedure for welding a butt joint on aluminum in the flat position.

Criteria:

Welding technique, weld appearance, and weld strength as evaluated by the instructor.

Procedure:

1. Refer to Exercise 4-12 for equipment set-up and adjustment.
2. Use the procedures detailed for welding mild steel in Exercise 4-3.

Exercise 4-14. Welding a Lap Joint on Aluminum in Horizontal Position Using the GTAW Process

Conditions:

Horizontal position
Refer to Exercise 4-12.
Two pieces of $^{1}/_{8}$" × 1$^{1}/_{2}$" × 6" aluminum

Performance:

The welder will demonstrate the correct procedure for welding a lap joint on aluminum in the horizontal position.

Criteria:

Welding technique, weld appearance, and weld strength as evaluated by the instructor.

Procedure:

1. Refer to Exercise 4-12 for equipment set-up and adjustment.
2. Use the procedures listed for welding mild steel in Exercise 4-4.

Exercise 4-15. Welding a T-Joint on Aluminum in Horizontal Position Using the GTAW Process

Conditions:

Horizontal position
Refer to Exercise 4-12.
Two pieces of $^1/_8$″ × $1^1/_2$″ × 6″

Performance:

The welder will demonstrate the correct procedure for welding a T-joint on aluminum in the horizontal position.

Criteria:

Welding technique, weld appearance, and weld strength as evaluated by the instructor.

Procedure:

1. Refer to Exercise 4-12 for equipment set-up and adjustment.
2. Use the procedures detailed for welding mild steel in Exercise 4-5.

Exercise 4-16. Running Beads on Mild Steel Using the GMAW Process

Conditions:

Flat position
DC reverse polarity
100–120 amps
19–21 volts
.035″ diameter wire, E-70S-3
Carbon dioxide shielding gas, 20 cfm
Electrode extension, $^1/_4$″–$^3/_8$″
$^3/_{16}$″–$^1/_4$″ × 4″ × 6″ mild steel plate

Performance:

Deposit beads on a mild steel plate using the GMAW process.

Criteria:

The beads should be approximately $^5/_{16}$″ wide and $^1/_8$″ high and consistent in width and straightness.

Procedure:

1. Set the wire feeder for the correct amperage setting as detailed in the text.
2. Adjust the voltage for the correct setting.

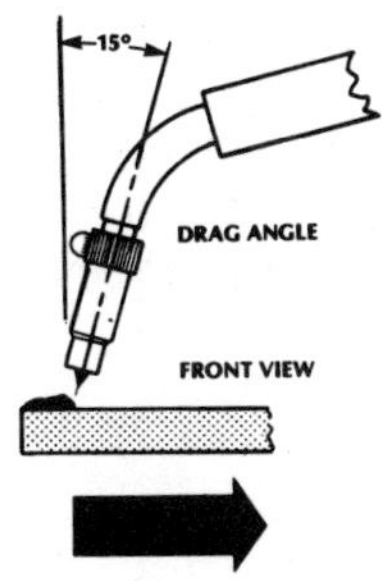

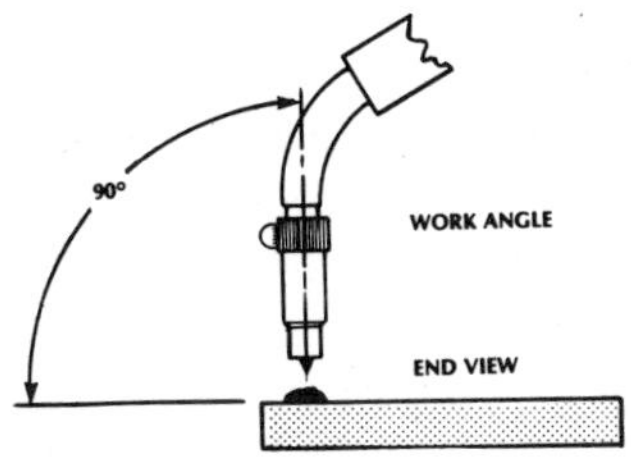

3. Use a 90° work angle and a 10°–15° drag angle.
4. Run a series of straight consistent beads approximately 3/8″ apart.

Exercise 4-17. Depositing a Metal Pad Using the GMAW Process

Conditions:

Flat position
Refer to Exercise 4-16.

Performance:

The welder will deposit successive overlapping beads to increase the thickness of the metal plate.

Criteria:

Each bead should overlap the previous bead in half and completely penetrate the previous bead and the base metal without voids in the weld metal applied.

Procedure:

1. Refer to Exercise 4-16 for equipment set-up and adjustment.
2. Deposit a bead 1/4″ from the edge, parallel to the length of the plate.
3. Use an 80°–90° work angle, and a 10°–15° drag angle to overlap the first bead in half.
4. Deposit consistent overlapping beads until the plate is covered.

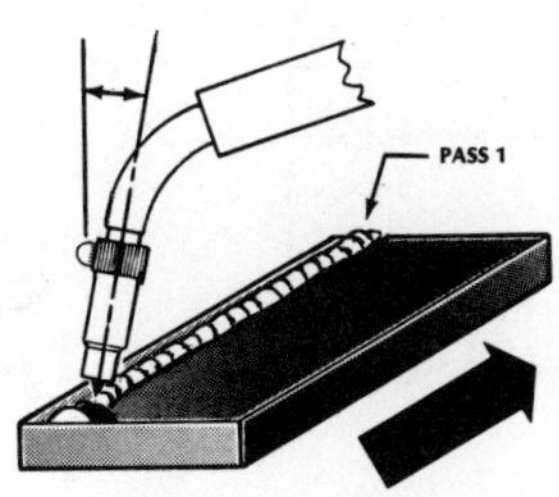

Exercise 4-18. Welding a Butt Joint on Mild Steel in Flat Position Using the GMAW Process

Conditions:

Flat position
Refer to Exercise 4-16.
Two 1/4″ × 1 1/2″ × 6″ mild steel plates

Performance:

The welder will demonstrate the correct procedure for welding a butt joint in the flat position.

Criteria:

Welding technique, weld appearance, and weld strength as evaluated by the instructor.

Procedure:

1. Refer to Exercise 4-16 for equipment set-up and adjustment.
2. Tack weld the two pieces to form a butt joint. Allow a $^{3}/_{32}$" root opening.
3. Use a 90° work angle and a 10° drag angle. A slight weaving motion may be used to control the puddle.

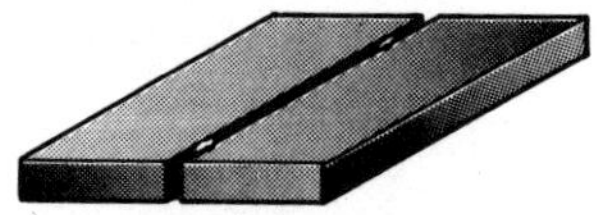

Exercise 4-19. Welding a Lap Joint on Mild Steel in Flat Position Using the GMAW Process

Conditions:

Flat position
Refer to Exercise 4-18.

Performance:

The welder will demonstrate the correct procedure for welding a lap joint in the flat position.

Criteria:

Welding technique, weld appearance, and weld strength as evaluated by the instructor.

Procedure:

1. Refer to Exercise 4-16 for equipment set-up and adjustment.
2. Tack weld the two pieces to form a lap joint.
3. Secure the tack welded piece in a positioner, or rest it against a firebrick to obtain flat position.
4. Use a 45° work angle and a 10°–15° drag angle with a slight weaving motion to deposit the bead.
5. The bead face should be flat to slightly convex.

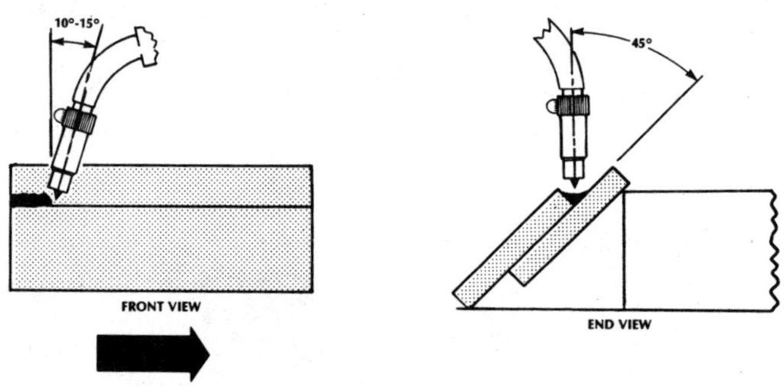

Exercise 4-20. Depositing Beads on Mild Steel in the Horizontal Position Using the GMAW Process

Conditions:

Horizontal position
Refer to Exercise 4-16.

Performance:

The welder will demonstrate the correct procedure for depositing beads in the horizontal position.

Criteria:

The beads should be approximately $^{5}/_{16}$" wide and $^{1}/_{8}$" high. The beads should be parallel to the length of the plate and free from undercut or overlap.

Procedure:

1. Refer to Exercise 4-16 for equipment set-up and adjustment.
2. Secure the plate in a positioner.
3. Use a 80° work angle and a 10°–15° drag angle.
4. Deposit a series of straight consistent beads approximately $^{3}/_{8}$″ apart.

Exercise 4-21. Welding a Multiple-Pass T-Joint in Horizontal Position Using the GMAW Process

Conditions:

Horizontal position
Refer to Exercise 4-16.
Two $^{1}/_{4}$″ × 2″ × 4″ mild steel plates

Performance:

The welder will demonstrate the correct procedure for welding a multiple-pass T-joint in horizontal position.

Criteria:

Welding technique, weld appearance, and weld strength as evaluated by the instructor.

Procedure:

1. Refer to Exercise 4-16 for equipment set-up and adjustment.
2. Tack weld the two pieces to form a T-joint.
3. Position the gun at a 45° work angle and a 10°–15° drag angle.
4. Deposit the first pass on both sides of the T-joint.

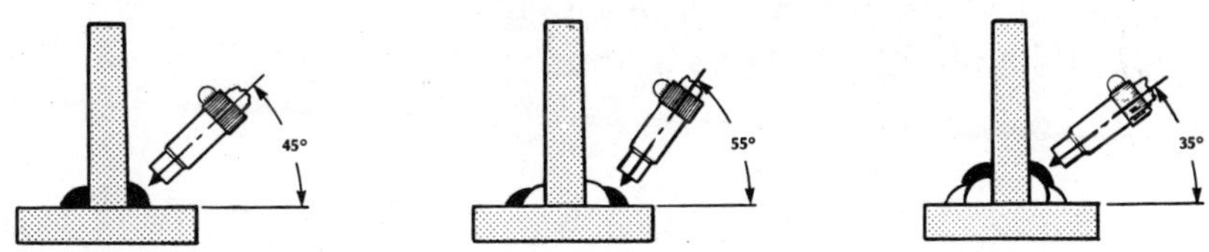

PASS 1 PASS 2 PASS 3

5. Position the gun at at 55° work angle and a 10°–15° drag angle.
6. Deposit the second pass, penetrating half of the first pass and the bottom plate on both sides of the joint.
7. Position the gun at a 35° work angle and a 10°–15° drag angle.
8. Deposit the third pass, penetrating half of the first pass and the top plate on both sides of the joint.

Exercise 4-22. Welding a Lap Joint on Mild Steel in Horizontal Position Using the GMAW Process

Conditions:

Horizontal position
Refer to Exercise 4-18.

Performance:

The welder will demonstrate the correct procedure for welding a lap joint in the horizontal position.

Criteria:

Welding technique, weld appearance, and weld strength as evaluated by the instructor.

Procedure:

1. Refer to Exercise 4-18 for equipment set-up and adjustment.
2. Tack weld the two pieces to form a lap joint.
3. Position the gun at a 45° work angle and a 10°–15° drag angle.
4. Deposit the weld with a slight weaving motion. Pause at the toes of the weld to avoid undercutting.
5. Repeat the weld on the other side of the joint.

Exercise 4-23. Welding a Butt Joint on Mild Steel in Horizontal Position Using the GMAW Process

Conditions:

Horizontal position
Refer to Exercise 4-18.

Performance:

The welder will demonstrate the correct procedure for welding a butt joint in the horizontal position.

Criteria:

Welding technique, weld appearance, and weld strength as evaluated by the instructor.

Procedure:

1. Refer to Exercise 4-16 for equipment set-up and adjustment.
2. Tack weld the two pieces to form a butt joint. Allow a $^{3}/_{32}$″ root opening.
3. Secure the tack welded piece in a positioner.
4. Use an 80° work angle and a 10°–15° drag angle.
5. Deposit a bead approximately $^{5}/_{16}$″ wide across the piece.

Exercise 4-24. Depositing Beads on Mild Steel in Vertical Position Using the GMAW Process

Conditions:

Vertical position
Refer to Exercise 4-21.

Performance:

The welder will demonstrate the correct procedure for depositing beads in both vertical up and vertical down positions.

Criteria:

The beads should be approximately $^{1}/_{4}$″ wide, $^{1}/_{8}$″ high, and parallel to the length of the plate.

Procedure:

1. Refer to Exercise 4-16 for equipment set-up and adjustment.
2. Secure the plate in a positioner.
3. Using the vertical down technique, position the gun with a 90° work angle and a 10°–15° drag angle.
4. Deposit a series of consistent beads approximately $^{3}/_{8}$″ apart.
5. On the other side of the plate, repeat the operation using the vertical up technique.

Exercises 4-25 and 4-26. Welding a Butt Joint on Mild Steel in Vertical Position Using the GMAW Process

Conditions:

Vertical position
Refer to Exercise 4-21.

Exercise 4-25. Vertical down
Exercise 4-26. Vertical up

Performance:

The welder will demonstrate the correct procedure for welding a butt joint in vertical position.

Criteria:

Welding technique, weld appearance, and weld strength as evaluated by the instructor.

Procedure:

1. Refer to Exercise 4-16 for equipment set-up and adjustment.
2. Tack weld the two pieces to form a butt joint. Allow a $^{3}/_{32}$" root opening.
3. Secure the tack welded piece in a positioner.
4. Start the weld at the top (Exercise 25), or at the bottom (Exercise 26).
5. Use a 90° work angle and a 15° drag angle with a slight weaving motion.

Exercise 4-27. Welding a T-Joint on Mild Steel in Vertical Position Using the GMAW Process

Conditions:

Vertical position
Refer to Exercise 4-21.

Performance:

The welder will demonstrate the correct procedure for welding a T-joint in vertical position.

Criteria:

Welding technique, weld appearance, and weld strength as evaluated by the instructor.

Procedure:

1. Refer to Exercise 4-16 for equipment set-up and adjustment.
2. Tack weld the two pieces to form a T-joint. Secure the tack welded piece in a positioner.
3. Start at the top and use a 45° work angle and a 10°–20° drag angle to deposit the bead.
4. Use a slight weaving motion, pausing at the toes to prevent undercutting.

Exercise 4-28. Welding a Single V-Butt Joint on Mild Steel in Vertical Position Using the GMAW Process

Conditions:

Vertical position
Refer to Exercise 4-21.

Performance:

The welder will demonstrate the correct procedure for welding a single V-butt joint in vertical position.

Criteria:

Welding technique, weld appearance, and weld strength as evaluated by the instructor.

Procedure:

1. Refer to Exercise 4-16 for equipment set-up and adjustment.
2. Bevel the two pieces to a 60° included angle with a $^3/_{32}$" root face.
3. Tack weld the two pieces to form a butt joint. Allow a $^3/_{32}$" root opening.
4. Secure the tack welded piece in a positioner.
5. Start the weld at the top using a 90° work angle and a 15° drag angle.
6. Use a weaving motion to deposit additional passes required to cover the groove faces.

Exercise 4-29. Welding a Single V-Butt Joint on Mild Steel in Overhead Position

Conditions:

Overhead position
Refer to Exercise 4-21.

Performance:

The welder will demonstrate the correct procedure for welding a single V-butt joint in overhead position.

Criteria:

Welding technique, weld appearance, and weld strength as evaluated by the instructor.

Procedure:

1. Refer to Exercise 4-28, Steps 1–4.
2. Use a 90° work angle and a 10°–15° drag angle to deposit the root pass.
3. Use a weaving motion to deposit additional passes required to cover the groove faces.

Exercise 4-30. Welding a Multiple-Pass T-Joint on Mild Steel in Overhead Position Using the GMAW Process

Conditions:

Overhead position
Refer to Exercise 4-21.

Performance:

The welder will demonstrate the correct procedure for welding a multiple-pass T-joint in the overhead position.

Criteria:

Welding technique, weld appearance, and weld strength as evaluated by the instructor.

Procedure:

1. Refer to Exercise 4-16 for equipment set-up and adjustment.
2. Tack weld the two pieces to form a T-joint.
3. Secure the tack welded piece in a positioner.
4. Deposit the first pass using a 45° work angle and a 5°–10° drag angle on both sides of the joint.
5. Deposit the second pass using a 50° work angle and a 5°–10° drag angle on both sides of the joint with a slight weaving motion.
6. Deposit the third pass using a 40° work angle and a 5°–10° drag angle on both sides of the joint with a slight weaving motion.

Exercise 4-31. Running Beads on Mild Steel Using the Flux Cored Arc Welding (FCAW) Process

Conditions:

Flat position
DC reverse polarity
390–410 amps
26–28 volts
$^{3}/_{32}$" diameter electrode, E70T-1
Carbon dioxide shielding gas, 40 cfm
Electrode extension, 1"–1$^{1}/_{2}$"
$^{1}/_{2}$"–1" × 4" × 6" mild steel plate

Performance:

The welder will deposit beads on a mild steel plate using the FCAW process.

Criteria:

The beads should be approximately $^{3}/_{4}$" wide, $^{1}/_{4}$" high, and parallel to the length of the plate.

Procedure:

1. Set the wire feeder for the correct amperage setting.
2. Adjust the voltage to the correct setting.
3. Use a 90° work angle and a 20°–30° drag angle.
4. Run a series of beads approximately $^{3}/_{8}$" apart. Remove the slag.

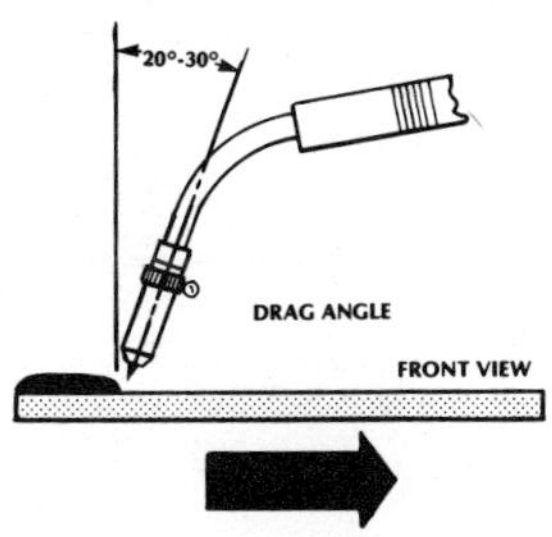

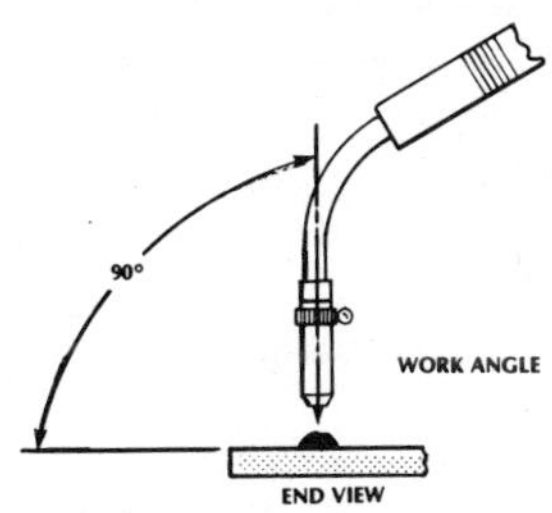

Exercise 4-32. Depositing a Metal Pad Using the FCAW Process

Conditions:

Flat position
Refer to Exercise 4-31.

Performance:

The welder will deposit successive overlapping beads to increase the thickness of the metal plate.

Criteria:

Each bead should overlap the previous bead in half, and completely penetrate the previous bead and the base metal without voids in the weld metal applied.

Procedure:

1. Refer to Exercise 4-30 for equipment set-up and adjustment.
2. Deposit a bead $^{1}/_{2}$" from the edge, parallel to the length of the plate. Remove the slag.
3. Use an 80° work angle and a 15° drag angle to overlap the first bead in half. Remove the slag.
4. Deposit consistent overlapping beads until the plate is covered.

Exercise 4-33. Welding a Lap Joint on Mild Steel in Flat Position Using the FCAW Process

Conditions:

Flat position
Refer to Exercise 4-31.
Two 3/4″ × 1″ × 2″ × 6″ mild steel plates

Performance:

The welder will demonstrate the correct procedure for welding a lap joint in flat position.

Criteria:

Welding technique, weld appearance, and weld strength as evaluated by the instructor.

Procedure:

1. Refer to Exercise 4-31 for equipment set-up and adjustment.
2. Tack weld the two pieces to form a lap joint. Remove the slag.
3. Secure the welded piece in a positioner or rest the piece against a firebrick to obtain flat position.
4. Use a 45° work angle and a 20° drag angle to deposit the first pass. Repeat this operation on the other side of the joint. Remove the slag.
5. Deposit additional passes required to fill the joint using a weaving motion. Pause at the toes of the weld to prevent undercutting.

Exercise 4-34. Welding a Multiple-Pass T-Joint in Horizontal Position Using the FCAW Process

Conditions:

Horizontal position
Refer to Exercise 4-33.

Performance:

The welder will demonstrate the correct procedure for welding a multiple-pass T-joint in horizontal position.

Criteria:

Welding technique, weld appearance, and weld strength as evaluated by the instructor.

Procedure:

1. Refer to Exercise 4-31 for equipment set-up and adjustment.
2. Tack weld the two pieces to form a T-joint. Remove the slag.

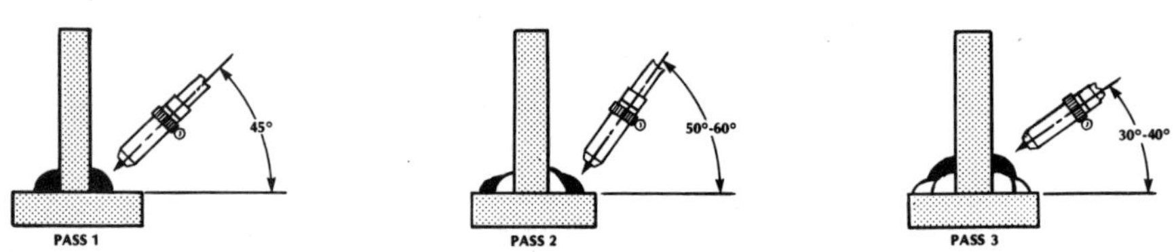

3. Use a 45° work angle and a 20° drag angle to deposit the first pass on both sides of the joint. Remove the slag.
4. Use a 50°–60° work angle and a 20° drag angle to deposit the second pass on both sides of the joint. The bead should penetrate half of the root pass and the base metal. Remove the slag.
5. Use a 30°–40° work angle and a 20° drag angle to deposit the third pass on both sides of the joint. The bead should penetrate half of the root pass and the base metal.

Exercise 4-35. Depositing Beads on Mild Steel in Horizontal Position Using the FCAW Process

Conditions:

Horizontal position
Refer to Exercise 4-31.

Performance:

The welder will demonstrate the correct procedure for depositing beads in the horizontal position.

Criteria:

The beads should be approximately 5/8″ wide, 3/16″–1/4″ high, and parallel to the length of the plate.

Procedure:

1. Refer to Exercise 4-31 for equipment set-up and adjustment.
2. Secure the plate in a positioner.
3. Use an 80° work angle and a 20° drag angle to deposit straight consistent beads approximately 1/2″ apart.

Exercise 4-36. Welding a T-Joint on Mild Steel in Vertical Position Using the FCAW Process

Conditions:

Vertical position
Refer to Exercise 4-33.

Performance:

The welder will demonstrate the correct procedure for welding a T-joint in the vertical down position.

Criteria:

Welding technique, weld appearance, and weld strength as evaluated by the instructor.

Procedure:

1. Refer to Exercise 4-31 for equipment set-up and adjustment.
2. Tack weld the two pieces to form a T-joint. Remove the slag.
3. Secure the tack welded piece in a positioner.
4. Start at the top and use a 45° work angle and a 20° drag angle to deposit the bead.
5. Use a slight weaving motion, pausing at the toes to avoid undercutting.

SECTION ACTIVITIES 5

Chapters 27–33 Special Welding Processes

1. List six types of resistance welding.

____

____

____

____

____

____

2. List three advantages that inertia welding has over conventional flash or butt welding.

____

____

____

3. List three uses of the ultrasonic welding process.

____

____

____

4. List five types of thermosetting plastics.

____

____

____

____

____

5. List five types of thermoplastics.

____

____

____

____

____

6. Define the following plastic welding techniques.

Hot Gas ______________________________

Heated Tool ______________________________

Induction ______________________________

Friction ______________________________

7. List and define the five major components of a robot welding system.

A. ______________________________

B. ______________________________

C. ______________________________

D. ______________________________

E. ______________________________

8. Use arrows to show the direction of joint movement of the five axes robot shown.

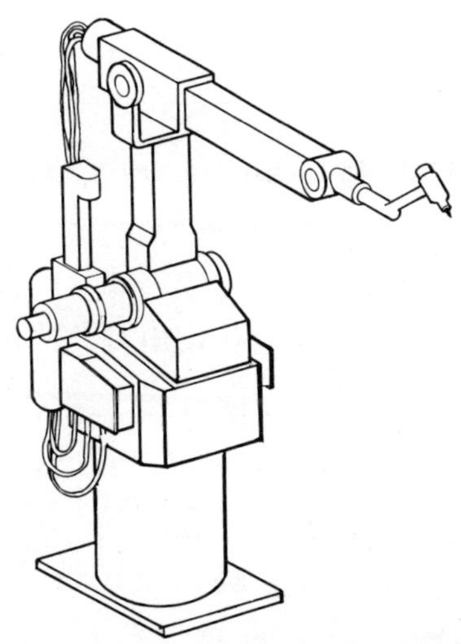

Exercises 5-1 through 5-3. Brazing Joints on Mild Steel Using the Oxyacetylene Flame

Conditions:

Flat position
Oxyacetylene welding equipment
Brazing rod (prefluxed)
$^{1}/_{16}'' \times 1^{1}/_{2}'' \times 5''$ mild steel plates

Performance:

The welder will demonstrate the correct procedure for brazing the following joints:

Exercise 5-1. Lap Joint
Exercise 5-2. Butt Joint with a Backing Plate
Exercise 5-3. Scarfed Butt Joint

Criteria:

Brazing technique, brazed joint appearance, and brazed joint strength as evaluated by the instructor.

Procedure:

1. Refer to Exercise 2-1 for oxyacetylene equipment and set-up.
2. Remove all dirt, grease, oil, and oxides from the pieces to be brazed.
3. Align the pieces in the correct position. Use jigs if necessary.
4. Adjust the oxyacetylene flame to neutral or slightly carburizing.
5. Preheat the pieces uniformly.
6. Touch the brazing rod to the preheated piece. As the filler metal melts, apply the rod with a forehand technique. Use a circular motion with the torch to distribute the filler metal and to form the bead.
7. Remove all flux residue.

Exercise 5-4. Braze Welding a Single V-Butt Joint on Mild Steel Using the Oxyacetylene Flame

Conditions:

Flat position
Refer to Exercises 5-1 through 5-3.
Two $^3/_{16}$" × 2" × 4" mild steel pieces

Performance:

The welder will demonstrate the correct procedure for braze welding a single V-butt joint.

Criteria:

Braze welding technique, braze weld joint appearance, and braze weld joint strength as evaluated by the instructor.

Procedure:

1. Bevel the pieces to a 60° included angle with a $^3/_{32}$" root face.
2. Refer to Exercises 5-1 through 5-3, Steps 1–5.
3. Hold the torch at the start of the joint. The metal should be heated until it turns red.
4. Melt a little brazing rod and spread it across the entire joint. This tinning operation prepares the joint to accept the filler metal necessary for joint strength.
5. Use a forehand technique and a circular torch motion to fill the groove of the joint.
6. The metal should be heated only enough to melt the filler metal. Do not overheat the base metal.

Exercises 5-5 and 5-6. Soldering Joints on Mild Steel Using a Soldering Copper

Conditions:

Flat position
Soldering copper
50/50 solder
Two 22-gauge, 2" × 6" mild steel pieces

Performance:

The welder will demonstrate the correct procedure for soldering the following joints:

Exercise 5-5. Seam Solder Butt Joint
Exercise 5-6. Sweat Solder Lap Joint

Criteria:

Soldering technique, soldered joint appearance, and soldered joint strength as evaluated by the instructor.

Procedure:

1. Remove all dirt, grease, oil, and oxides from the pieces to be soldered.
2. Align the pieces in the correct position. Use jigs if necessary.
3. Apply flux to the areas to be tinned on the joints.
4. Prepare and tin the soldering copper.

SEAM SOLDERING

5. Tack the fluxed pieces together in several places (seam soldering).
6. Hold the soldering copper on the joint until the flux begins to sizzle. Begin applying solder behind the copper while carefully advancing the copper across the joint. The solder should melt from the heat of the metal, not the copper (seam soldering).
7. Apply a uniform coating of solder on the surfaces to be joined (sweat soldering).
8. Place the surfaces together with the soldered sides in contact (sweat soldering).
9. Place the flat side of the copper on the seam. As the solder begins to melt and flow out of the joint, draw the copper slowly across the joint. Use a punch to hold the pieces in contact until sufficient cooling has taken place and the solder has solidified (sweat soldering).

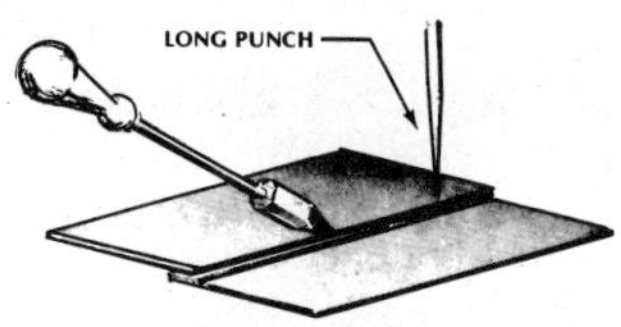

SWEAT SOLDERING

Exercise 5-7. Hardfacing Mild Steel Using the Shielded Metal-Arc Welding (SMAW) Process

Conditions:

Flat position
Refer to Exercise 3-25.
Hardfacing electrode
Worn part or steel plate as provided by the instructor

Performance:

The welder will demonstrate the correct procedure for hardfacing the mild steel provided by the instructor.

Criteria:

Hardfacing technique, hardfacing appearance, and hardfacing strength as evaluated by the instructor.

Procedure:

1. Obtain a metal piece to be hardfaced from the instructor.
2. Clean the surface by removing rust, scale, or other foreign matter.
3. Use only enough amperage to maintain the arc necessary to heat the base metal.
4. Arrange the work for flat position.
5. Maintain a long arc length. Avoid touching the electrode to the area being hardfaced.
6. A weaving motion may be used in areas where a thin deposit is required.
7. Remove all slag from the surface before depositing additional layers.

Exercises 5-8 through 5-11. Welding Thin-Wall Pipe Using the SMAW Process

Conditions:

Exercise 5-8. Test Position 1G
Exercise 5-9. Test Position 2G
Exercise 5-10. Test Position 5G
Exercise 5-11. Test Position 6G

E-6010 $^{1}/_{8}$″ DC reverse polarity
Thin-wall pipe

Performance:

The welder will demonstrate the correct procedure for welding thin-wall pipe in the four pipe weld test positions.

Criteria:

Welding technique, weld appearance, and weld strength as evaluated by the instructor.

Procedure:

1. Obtain thin-wall ($^{1}/_{8}$″ to $^{5}/_{16}$″) pipe lengths from the instructor.

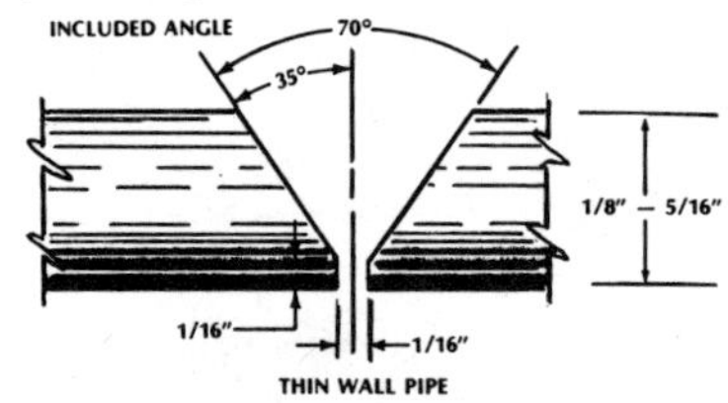

2. Prepare the joints as shown.

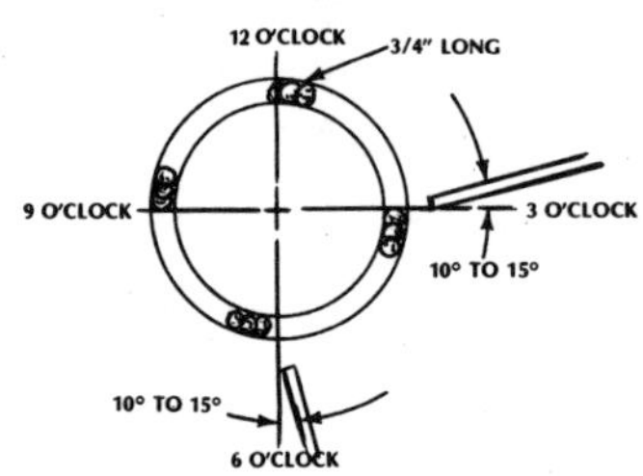

3. Tack weld the pipe at 3, 6, 9, and 12 o'clock locations with a $^{3}/_{32}$″ root opening.
4. Using the downhill technique, weld the pipe in the four pipe weld test positions.

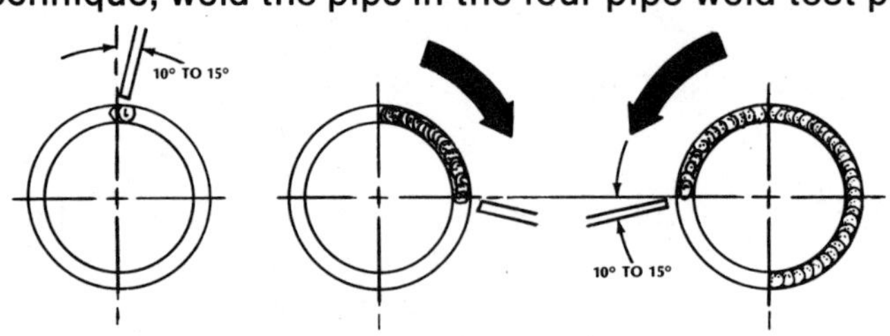

Exercises 5-12 through 5-15. Welding Thick-Wall Pipe Using the SMAW Process

Conditions:

Exercise 5-12. Test Position 1G
Exercise 5-13. Test Position 2G
Exercise 5-14. Test Position 5G
Exercise 5-15. Test Position 6G

E-6010 $^{1}/_{8}$″ DC reverse polarity
Thick-wall pipe

Performance:

The welder will demonstrate the correct procedure for welding thick-wall pipe in the four pipe weld test positions.

Criteria:

Welding technique, weld appearance, and weld strength as evaluated by the instructor.

Procedure:

1. Obtain thick-wall (over $^5/_{16}$″) pipe lengths from the instructor.
2. Prepare the joints as shown.

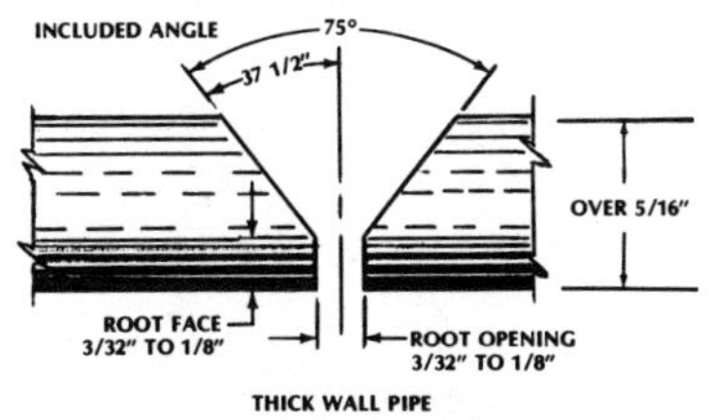

3. Tack weld the pipe at 3, 6, 9, and 12 o'clock locations with a root opening of approximately $^3/_{32}$″ to $^1/_8$″.
4. Using the uphill technique, weld the pipe in the four pipe weld test positions.

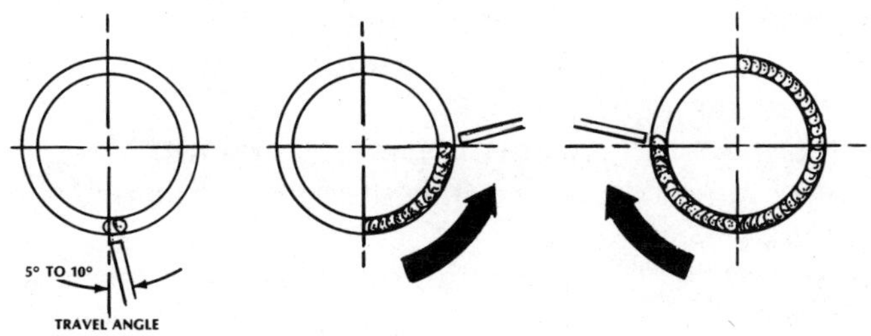

Exercises 5-16 through 5-19. Welding Thin-Wall Pipe Using the Gas Metal-Arc Welding Process

Conditions:

Exercise 5-16. Test Position 1G
Exercise 5-17. Test Position 2G
Exercise 5-18. Test Position 5G
Exercise 5-19. Test Position 6G

.035″ diameter wire, E-70S-3
Refer to Exercise 4-16.
Thin-wall pipe

Performance:

The welder will demonstrate the correct procedure for welding thin-wall pipe in the four pipe weld test positions.

Criteria:

Welding technique, weld appearance, and weld strength as evaluated by the instructor.

Procedure:

1. Obtain thin-wall ($^1/_8$″ to $^5/_{16}$″) pipe lengths from the instructor.
2. Prepare the joints as shown in Exercises 5-8 through 5-11.
3. Tack weld the pipe at 3, 6, 9, 12 o'clock locations with a $^3/_{32}$″ root opening.
4. Use the downhill technique to weld the pipe in the four pipe weld test positions.
5. Repeat the exercise using the vertical up technique.

Exercises 5-20 through 5-24. Flame Cutting Using the Oxyacetylene Cutting Torch

Conditions:

Flat position
Oxyacetylene cutting equipment
$^1/_2$″ steel plate

Performance:

The welder will demonstrate the correct procedure for cutting:

Exercise 5-20. Straight Cuts
Exercise 5-21. Curves and Irregular Shapes
Exercise 5-22. Piercing Holes
Exercise 5-23. Bevels
Exercise 5-24. Round-Bar Steel

Criteria:

Cutting technique, cut area appearance, and cut accuracy as evaluated by the instructor.

Procedure:

1. Obtain pieces of mild steel plate from the instructor. Use soapstone to mark the cutting areas.
2. Mount the correct size cutting tip in the cutting torch.
3. Adjust the gas pressures as recommended.
4. Light the torch and adjust the preheat flame to neutral.
5. With the oxygen lever turned on, adjust the cutting flame so that the preheating cones are burning with a neutral flame.
6. Practice cutting straight cuts, curves, and bevels; piercing holes and round-bar steel.

Exercise 5-25. Cutting Mild Steel Using the Air Carbon-Arc Cutting Process

Conditions:

Flat, horizontal, and vertical positions
Air carbon-arc equipment
Compressed air
Rectifier type welding machine
$^1/_2$″ to 1″ mild steel plate

Performance:

The welder will demonstrate the correct procedure for gouging, cutting, washing, and beveling mild steel using the air carbon-arc cutting process.

Criteria:

Cutting technique, cut area appearance, and cut accuracy as evaluated by the instructor.

Procedure:

1. Obtain pieces of mild steel plate from the instructor.
2. Adjust the welding machine to the correct amperage and polarity.
3. Set the compressed air to the pressure required.
4. Grip the electrode so that a maximum of 6″ extends from the electrode holder.
5. Check to see that the air jet orifices are positioned under the electrode.
6. Adjust the push angle for cutting, washing, and beveling operations.
7. Practice cutting, washing, and beveling in the flat, horizontal, and vertical positions.

SECTION ACTIVITIES 6

Chapters 34-36 Supplementary Welding Data

1. Prepare three face and root bend specimens, and use a guided bend tester to check for the quality of fusion and degree of weld penetration.

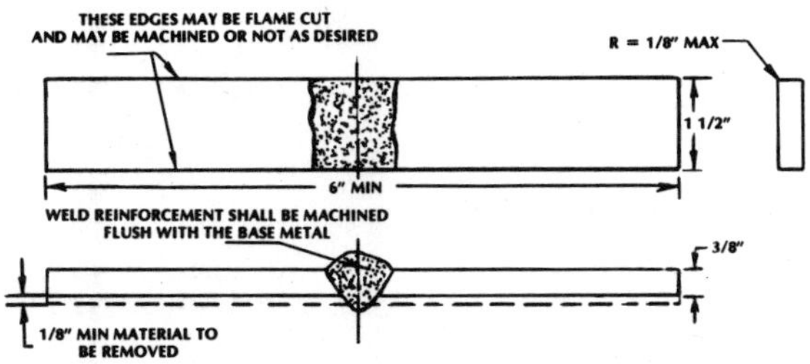

FACE AND ROOT BEND SPECIMEN FOR PLATE

2. Prepare four multiple-pass T-joints. Use the visual examination method to check for the following external weld defects. Mark the checklist "OK" if no defect is present. Place a check by the defect that is present.

WELD DEFECTS PRESENT

	Specimen 1	Specimen 2	Specimen 3	Specimen 4
Cracks				
Undercut				
Overlap				
Slag Inclusions				
Inadequate Penetration				
Improper Bead Size				
Improper Bead Contour				

3. Draw the welding symbol that describes the weld shown.

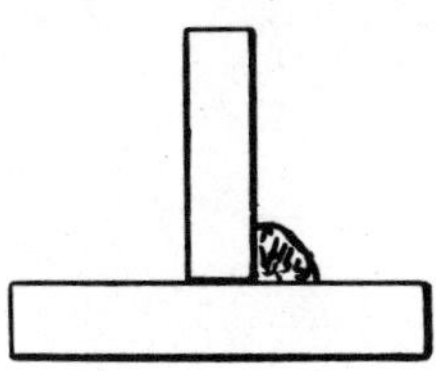
A

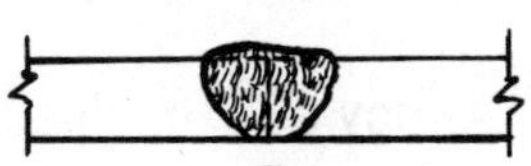
B

C

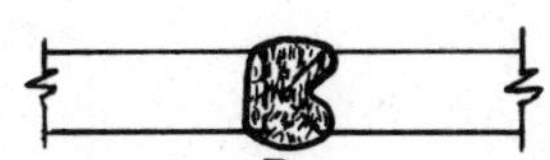
D

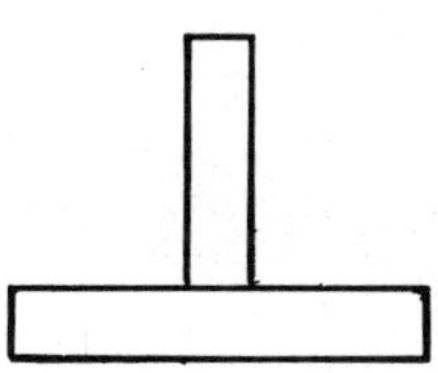
A

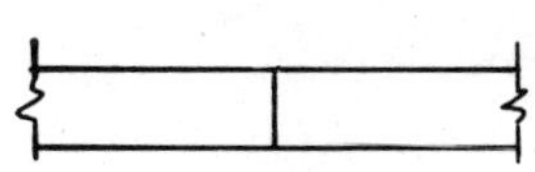
B

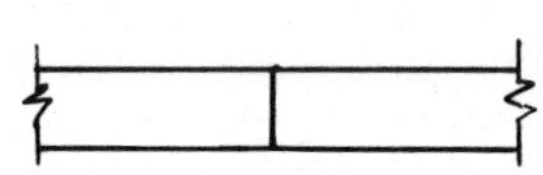
C

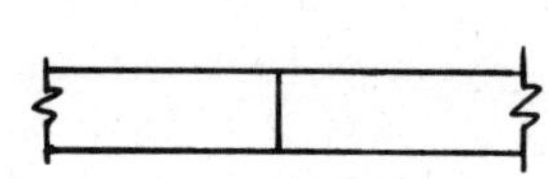
D

4. Perform the following groove welds in flat, horizontal, vertical, and overhead positions using the shielded metal-arc welding process. Check off when successfully completed.

AWS Qualification Designations for Welding Competencies

1G Flat position groove weld
2G Horizontal position groove weld
3G Vertical position groove weld
4G Overhead position groove weld

Position	E-6010	E-6013	E-7018
1G			
2G			
3G			
4G			

5. Perform the following pipe welds in the AWS test positions listed. Check off when successfully completed.

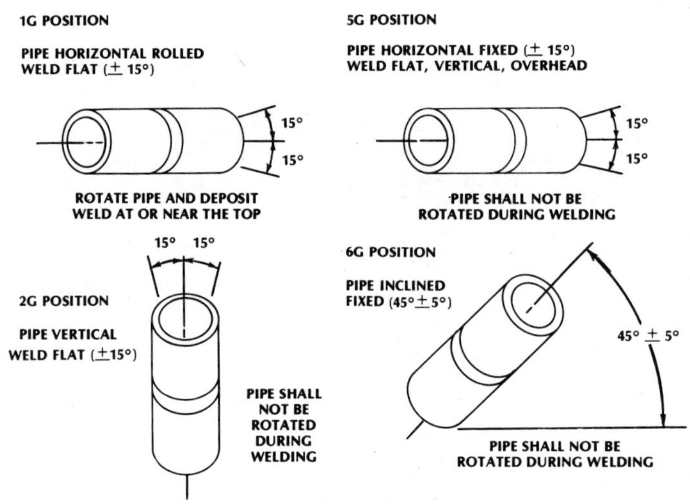

Position	SMAW E-6010	SMAW E-7018	GMAW E-70S-3
1G			
2G			
5G			
6G			

SECTION EXAM 1

Name ____________________

Class ____________ Date ____________

Score ____________

Chapters 1-4 Introduction to Welding

True-False

T F **1.** Welding was first used in 2000 B.C.

T F **2.** Production work is one of the job tasks for a semi-skilled welder.

T F **3.** One advantage of the oxyacetylene welding process is the mobility of the necessary equipment.

T F **4.** Shielded metal-arc welding was developed before gas metal-arc welding.

T F **5.** Good safety habits must be practiced constantly.

T F **6.** A respirator should be used when welding metals that give off toxic fumes.

T F **7.** Hollow castings should not be welded without proper venting.

T F **8.** All injuries should be reported immediately to the supervisor.

T F **9.** A metal must be properly annealed to increase its hardness.

T F **10.** The grain size of metals is affected by the application of heat.

T F **11.** A correct preheat temperature can be determined by using a temperature indicating crayon.

T F **12.** Cryogenic properties refer to characteristics of metals when they are subjected to very high temperatures.

T F **13.** Nickel as an alloying element increases the ductility of steel.

T F **14.** Postweld heat treatment involves heating the metal after the welding operation is complete.

T F **15.** Tack welds are used to control expansion and contraction forces on long seams.

T F **16.** Chill blocks are used to hold the weld pieces in a fixed position while welding is taking place.

T F **17.** Metal contracts when heated.

T F **18.** Cyaniding is a type of case hardening process.

T F **19.** A crater in the weld is a depression in the base metal made by the welding heat source.

T F **20.** The face of a weld is its exposed surface bounded by its toes.

Multiple Choice

__________ **1.** The __________ welding process involves two pieces of metal placed between electrodes, which become conductors for low voltage and high amperage current.

A. shielded metal-arc
B. oxyacetylene
C. gas tungsten-arc
D. resistance
E. none of the above

__________ **2.** The __________ welding process uses an intense flame to generate the heat necessary to join metals.

A. gas metal-arc
B. gas tungsten-arc
C. oxyacetylene
D. resistance
E. none of the above

_______ **3.** The _______ welding process is used primarily for production welding.
A. shielded metal-arc
B. resistance
C. gas tungsten-arc
D. gas metal-arc
E. none of the above

_______ **4.** _______ welding was the first type of welding used by man.
A. Oxyacetylene
B. Carbon-arc
C. Forge
D. Stick
E. none of the above

_______ **5.** When welding _______, special precautions must be taken because of the toxic fumes emitted.
A. brass
B. zinc
C. lead
D. galvanized metals
E. all of the above

_______ **6.** Before welding on a container that has held unknown substances, the _______ to prevent possible injury.
A. hot chemical solution cleaning method must be used
B. exact substance which was held by the container must be determined
C. steam cleaning method must be used
D. chemical cleaning method with trisodium phosphate must be used
E. none of the above

_______ **7.** When ventilating a closed container before welding, _______ must be used.
A. acetylene
B. mapp gas
C. propane
D. oxygen
E. none of the above

_______ **8.** _______ properties of metals are those that include corrosion, oxidation, and reduction.
A. Physical
B. Chemical
C. Mechanical
D. Critical
E. none of the above

_______ **9.** _______ properties refer to the behavior of metals under applied loads.
A. Physical
B. Chemical
C. Mechanical
D. Ferritic
E. all of the above

_______ **10.** The last point to which a metal may be stretched and still return to its original condition is called its _______.
A. stress
B. modulus of elasticity
C. elastic limit
D. expansion strength
E. none of the above

________ **11.** A(n) ________ is a type of weld.
A. overhead
B. lap
C. fillet
D. corner
E. none of the above

________ **12.** A welding position commonly used is the ________ position.
A. flat
B. horizontal
C. vertical
D. overhead
E. all of the above

________ **13.** The amount of weld metal deposited above the surface of the pieces being joined is the ________.
A. root
B. throat
C. root face
D. penetration
E. reinforcement

________ **14.** The shape within the deposited bead caused by the heat of the welding source is the ________ of the weld.
A. penetration
B. toe
C. crown
D. ripple
E. none of the above

________ **15.** The ________ joint does not require edge preparation.
A. square butt
B. square T
C. single lap
D. double lap
E. all of the above

________ **16.** The distance through the center of the weld from face to root is the ________ of the weld.
A. throat
B. face
C. toe
D. root opening
E. all of the above

________ **17.** The point of the fillet weld triangle opposite the face of the weld is the ________ of the weld.
A. face
B. root
C. root opening
D. weld legs
E. none of the above

________ **18.** A fillet weld is used in a ________.
A. butt joint
B. groove joint
C. plug weld
D. T-joint
E. all of the above

______ **19.** The points where the weld metal meets the base metal in a fillet weld are the ______________ of the weld.
A. ripples
B. craters
C. roots
D. throats
E. toes

______ **20.** The distance from toe to toe across the face of a fillet weld is the ______ of the weld.
A. throat
B. face
C. root
D. toe
E. none of the above

______ **21.** Of the main types of welding positions, the ______ is the most difficult to perform.
A. flat
B. underhand
C. vertical
D. overhead
E. horizontal

Identify the mechanical properties shown.

______ **22.**
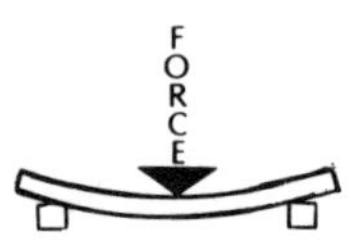

A. elasticity
B. stress
C. strain
D. elastic limit
E. none of the above

______ **23.**
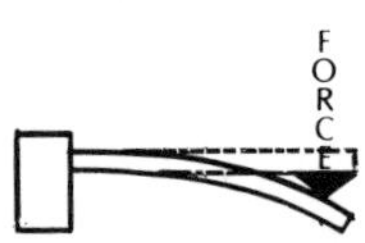

A. elasticity
B. stress
C. strain
D. impact strength
E. none of the above

______ **24.**

A. stress
B. tensile strength
C. compressive strength
D. elastic limit
E. none of the above

______ **25.**
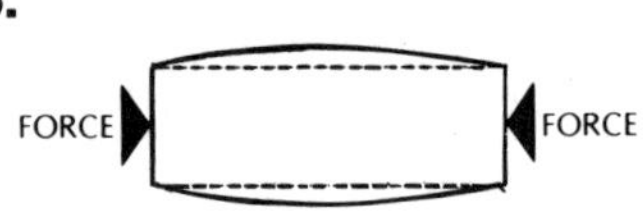

A. strain
B. shear strength
C. tensile strength
D. bending strength
E. none of the above

______ **26.**
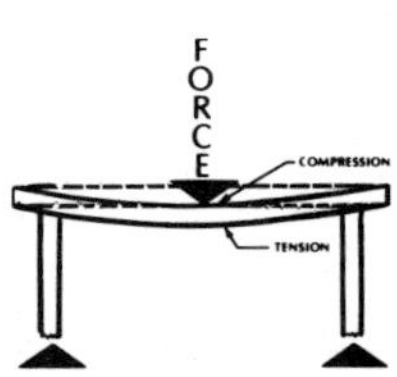

A. malleability strength
B. toughness
C. compressive strength
D. bending strength
E. none of the above

SECTION EXAM 2

Name ____________________

Class __________ Date __________

Score __________

Chapters 5-11 Oxyacetylene Welding — OAW

True-False

T F **1.** Oxygen cylinders are charged at a pressure of 2,200 psi.

T F **2.** A flash arrestor is used to ignite the welding torch.

T F **3.** The connecting nut used to mount the acetylene hose on the welding torch is notched to indicate left-hand threads.

T F **4.** Acetylene becomes unstable at 250 psi.

T F **5.** Welding goggles with a number 5 shade can be used for most oxyacetylene welding operations.

T F **6.** Needle valves on the welding torch control the gas pressure from the cylinders.

T F **7.** Purging the welding hoses involves removal of residual gases.

T F **8.** The thickness of the metal being welded determines the size of the welding torch tip to use.

T F **9.** A carburizing flame has excess oxygen.

T F **10.** When opening the oxygen and acetylene cylinder valves, stand to the side to avoid injury in case a regulator explodes.

T F **11.** Check valves are installed in line to prevent a flashback from reaching the oxygen and acetylene cylinders.

T F **12.** A torch is ignited with the tip facing upward.

T F **13.** A popping noise usually indicates an insufficient flow of gases to the torch tip.

T F **14.** The cylinder valve regulates gas flow to the torch.

T F **15.** The oxygen and acetylene cylinders are chained to a stationary object to prevent tipping.

T F **16.** An excessive amount of heat applied results in burn-through in the base metal.

T F **17.** When welding a T-joint in the flat position, the welding torch is held at a 30° angle to the work.

T F **18.** The diameter of the filler rod used should be twice the thickness of the metal when oxyacetylene welding mild steel.

T F **19.** The welding torch is held approximately ½″ from the work when carrying a puddle without a filler rod.

T F **20.** Welding in the flat position is easier and faster than in other welding positions.

T F **21.** If the weld puddle becomes too large when welding in positions other than the flat position, more heat should be directed to the weld area.

T F **22.** When horizontal welding a butt joint, more heat should be directed to the lower plate.

T F **23.** When welding heavy plates thicker than ¼″, edge preparation is not required.

T F **24.** The guided bend tester can be used to test the strength of a butt weld on heavy plate.

T F **25.** Cast iron must be preheated to a dull red to prevent stresses from contraction and expansion caused by stresses when the piece is welded.

T F **26.** Cast iron should be cooled rapidly to avoid distortion caused from the heat of welding.

T F **27.** When heated to its melting point, aluminum does not change color.

T F **28.** Carpenter's chalk can be used to determine the correct preheat temperature of aluminum.

T F **29.** The edges of thick gauge aluminum must be notched before welding to ensure adèquate penetration.

T F **30.** Aluminum must be cleaned thoroughly before it is welded to prevent impurities from weakening the weld.

Multiple Choice

_______ **1.** Which of the following statements about acetylene is not true?
A. It is usually transported in a red hose.
B. If compressed to more than 15 psi it becomes unstable.
C. It is less sensitive compared to mapp gas.
D. It is ignited first when lighting the torch.
E. none of the above

_______ **2.** To move an oxygen or acetylene cylinder, the cylinder should be carefully _______ to the desired location.
A. lifted
B. rolled on its side
C. lifted by the protector cap
D. rolled on its bottom edge
E. none of the above

_______ **3.** When opening the oxygen cylinder valve, it is necessary to _______.
A. use a wrench
B. open the valve one-half turn
C. open the valve all the way
D. remove the hand wheel
E. none of the above

_______ **4.** The difference between acetylene and oxygen is determined by comparing _______.
A. red hose color to green hose color
B. left-hand threads to right-hand threads
C. hose connecting nuts without notches to those with notches
D. AC torch markings to OX torch markings
E. all of the above

_______ **5.** An acetylene leak is usually easier to detect than an oxygen leak because _______.
A. of a quicker drop in regulator pressure
B. of its distinctive smell
C. it has less pressure in the cylinder
D. the gauge is easier to read
E. none of the above

_______ **6.** When welding in the flat position, the heat of the torch should be concentrated _______.
A. on the filler rod
B. on the base metal
C. around the weld area
D. all of the above
E. none of the above

_______ **7.** Which of the following statements about running beads in the flat position is not true?
A. A semicircular or circular torch movement is used.
B. Filler rod is dipped into the middle of the molten puddle.
C. The filler rod will stick if touched to cold metal outside of the puddle.
D. The torch is held at a 45° angle.
E. none of the above

______ **8.** When welding a butt joint in the vertical position, more of the flame should be directed to the ______.
- A. weld puddle
- B. area surrounding the weld
- C. filler rod
- D. top of the plate
- E. all of the above

______ **9.** Metal thicker than $^1/_8''$ is designated as ______.
- A. sheet
- B. clad steel
- C. coated
- D. plate
- E. none of the above

______ **10.** When welding gray cast iron, which of the following are correct procedures?
- A. Preheat the metal to a dull red.
- B. Bevel the edges to a 90° included angle.
- C. Use the correct recommended flux.
- D. Make sure that the filler rod possesses the same properties as the base metal.
- E. all of the above

______ **11.** Which of the following is not true about welding aluminum with the oxyacetylene welding process?
- A. Preheat the metal to approximately 300°-500°F.
- B. Use a tip larger than the tip used for welding mild steel.
- C. Clean the metal thoroughly before welding.
- D. Apply heat to the lower plate when welding a lap joint.
- E. none of the above

Matching

______ **1.** A ______ prevents the possibility of a flashback from reaching the manifold system.

______ **2.** The ______ controls flow of gases at the torch.

______ **3.** A ______ is used to ignite the flame.

______ **4.** The ______ converts cylinder gas pressure into working pressure.

______ **5.** The ______ shuts off gas flow from the cylinder.

- A. striker
- B. check valve
- C. regulator
- D. cylinder valve
- E. needle valve

______ **6.** When burning with low gas flow ______ produces black soot.

______ **7.** A balanced flame used for most welding applications is the ______ flame.

______ **8.** The ______ flame has an excess flow of oxygen.

______ **9.** After the flame is ignited, ______ is added to produce a neutral flame.

______ **10.** The ______ flame has an extended feather beyond the inner cone.

- A. oxygen
- B. acetylene
- C. carburizing
- D. oxidizing
- E. neutral

SECTION EXAM 3

Name ____________________

Class ____________ Date ____________

Score ____________

Chapters 12-23 Shielded Metal-Arc Welding — SMAW

True-False

T F **1.** DC reverse polarity is used with electrodes that function best with the electrons flowing into the work.

T F **2.** Resistance occurs when material in the conductor opposes the passage of electrical current.

T F **3.** The transformer type of welding machine is used with DC current only.

T F **4.** Safety glasses are not necessary if a helmet is worn when shielded metal-arc welding.

T F **5.** High voltage and low current are used in the SMAW process.

T F **6.** A chipping hammer is used to remove slag from the weld.

T F **7.** The ground cable is positive when welding with direct current straight polarity.

T F **8.** The constant current welding machine is used primarily for gas metal-arc applications.

T F **9.** Alternating current in the United States is rated at 60 cycles per second.

T F **10.** A voltmeter is used to measure amperage.

T F **11.** The E-6013 electrode is classified as a fill-freeze electrode.

T F **12.** The position of the weld is a factor in determining the type of electrode used.

T F **13.** Fast-fill electrodes are used for welding in the overhead position.

T F **14.** The coating on the electrode provides a shield which protects the molten metal from atmospheric contamination.

T F **15.** DC-EN on the electrode indicates that alternating current must be used.

T F **16.** The first two digits in the AWS numerical electrode classification refer to the tensile strength of the wire of the electrode.

T F **17.** Before striking the arc, the base metal must be free of any dirt, grease, or oil.

T F **18.** The electrode holder is used to hold the coated end of the electrode.

T F **19.** If an electrode sticks to the work, the amperage should be increased slightly.

T F **20.** An arc length that is too long results in excess spatter and poor penetration.

T F **21.** The weld crater is formed as the arc comes in contact with the base metal.

T F **22.** A whipping motion can be used when welding with an E-6013 electrode.

T F **23.** Undercutting can be prevented by lowering the amperage and changing the work angle of the electrode.

T F **24.** A multiple-pass weld requires at least two or more passes to deposit the required amount of weld metal.

T F **25.** A weaving motion is often used when depositing a cover pass.

T F **26.** Welding is simplified if the work is positioned in the flat position.

T F **27.** A lap joint requires edge preparation.

T F **28.** Welding speed can be increased if welding is done in the flat position.

T F **29.** Welding in the vertical down position allows greater penetration than welding in the vertical up position.

T F **30.** The E-7018 electrode does not require a whipping motion.

T F **31.** Welding in the vertical position requires a longer arc length than welding in the flat position.

T F **32.** The filler pass is deposited after the root pass and before the cover pass.

T F **33.** Welding in the overhead position requires that the amperage be reduced.

T F **34.** A positioner is used to secure the weld pieces when practicing welds in the overhead position.

T F **35.** The correct preheat temperature is necessary to prevent cracks caused from stress when welding cast iron.

T F **36.** Cracks in cast iron to be welded do not require edge preparation.

T F **37.** High-carbon steels are more difficult to weld than low-carbon weld.

T F **38.** Alloy steels require special treatment to obtain a satisfactory weld.

T F **39.** Pure aluminum is very stiff and brittle.

T F **40.** Aluminum should be preheated to a dull red before it is welded.

Multiple Choice

_______ **1.** Electrical current flowing in one direction only is _______.
- A. direct current
- B. alternating current
- C. static current
- D. dynamic current
- E. none of the above

_______ **2.** _______ is the force that causes current to move.
- A. Resistance
- B. Amperage
- C. Conductor
- D. Voltage
- E. none of the above

_______ **3.** _______ is electricity in motion in an electrical circuit.
- A. Constant current
- B. Resistance
- C. Voltage
- D. Dynamic electricity
- E. Constant electricity

_______ **4.** _______ voltage is produced when no welding is being done.
- A. Variable
- B. Reverse polarity
- C. Open circuit
- D. Short
- E. none of the above

_______ **5.** A generator type of welding machine produces _______ for welding.
- A. reverse polarity voltage
- B. alternating current
- C. direct current
- D. alternating current straight polarity
- E. all of the above

_______ **6.** Shielded metal-arc welding should be done using a number _______ or greater shade.
- A. 3
- B. 10
- C. 4
- D. 5
- E. none of the above

________ **7.** The ________ measures the amount of current flowing in the circuit.
A. voltmeter
B. ammeter
C. altimeter
D. resistance gauge
E. none of the above

________ **8.** When welding using DC straight polarity the electrode is ________.
A. positive
B. negative
C. neutral
D. alternating current
E. all of the above

________ **9.** The last digit on an electrode using the AWS numerical electrode classification refers to ________.
A. tensile strength
B. electric welding
C. welding position
D. special manufacturer's characteristics
E. all of the above

________ **10.** A(n) ________ electrode is best suited for vertical and overhead welding.
A. fast-fill
B. fast-freeze
C. fill-freeze
D. iron powder
E. all of the above

________ **11.** An ________ is recommended for flat position and horizontal fillets only.
A. E-6010
B. E-6013
C. E-7018
D. E-7024
E. none of the above

________ **12.** An ________ produces a flat or concave bead using DC reverse current.
A. E-6010
B. E-6012
C. E-6013
D. E-7024
E. all of the above

________ **13.** ________ covers the weld metal as it cools to prevent atmospheric contamination.
A. Wire
B. Slag
C. The arc length
D. The crater
E. none of the above

________ **14.** An ________ can be used with DC reverse polarity.
A. E-6010
B. E-6011
C. E-6013
D. E-7018
E. all of the above

________ **15.** The ________ is the rate at which an electrode is moved across the weld area.
A. direction of travel
B. arc length
C. electrode angle
D. speed of travel
E. all of the above

________ 16. The ________ is the distance between the electrode and the work.
A. arc length
B. work angle
C. electrode gap
D. current length
E. none of the above

________ 17. A ________ motion is used to control the heat of the weld by withdrawing the arc in and out of the weld crater.
A. backstep
B. forehand
C. whipping
D. backhand
E. none of the above

________ 18. ________ is the process of building up worn surfaces by depositing successive weld beads.
A. Whipping
B. Undercutting
C. Padding
D. Overlapping
E. all of the above

________ 19. ________ occurs when the amperage is set too high.
A. Overlapping
B. Padding
C. Weaving
D. Undercutting
E. none of the above

________ 20. ________ is a technique used to increase the width of the bead.
A. Restarting
B. Weaving
C. Padding
D. Surfacing
E. all of the above

________ 21. Weld metal deposited with the amperage set too low results in ________.
A. undercut
B. overlap
C. excessive spatter
D. porosity
E. none of the above

________ 22. ________ are used to hold the weld pieces in position before the root pass is deposited.
A. Filler passes
B. Intermittent welds
C. Backstep welds
D. Tack welds
E. all of the above

________ 23. A ________° work angle is used when depositing the first pass of a multiple-pass T-fillet weld in the flat position.
A. 15
B. 30
C. 45
D. 60
E. none of the above

_______ **24.** When welding plates of different thicknesses, more heat should be directed to _______ plate.
A. the thinner
B. the thicker
C. either
D. all of the above
E. none of the above

_______ **25.** A slight _______ motion is used when depositing beads vertical down using an E-6012 or E-6013 electrode.
A. weaving
B. drag
C. figure 8
D. crescent
E. all of the above

_______ **26.** A _______ pass is deposited to achieve a smooth weld appearance.
A. root
B. filler
C. cover
D. hot
E. none of the above

_______ **27.** The _______ position is not one of the four main welding positions.
A. forehand
B. vertical
C. overhead
D. horizontal
E. all of the above

_______ **28.** To prevent the spread of a crack in casting, _______ at the end of the crack.
A. V the area
B. peen the casting before welding
C. drill a 1/8″ hole
D. hold the torch
E. none of the above

_______ **29.** Most cast iron electrodes are designed to be used with _______.
A. DC straight polarity
B. DC reverse polarity and DC straight polarity
C. AC and DC reverse polarity
D. AC straight polarity
E. none of the above

_______ **30.** Of the carbon steel group, _______ -carbon steel is the easiest to weld.
A. low
B. medium
C. high
D. very high
E. all of the above

_______ **31.** When welding high-carbon steel, _______ is required to ensure a sound weld.
A. preheating
B. postweld heat treatment
C. using electrodes that match the properties of the base metal
D. all of the above
E. none of the above

_______ 32. A _______ can be used to determine the correct preheating procedures.
A. guided bend test
B. pair of pliers
C. chill plate
D. clip test
E. none of the above

_______ 33. An alloy steel is a steel mixed with _______ or more additional elements.
A. one
B. two
C. three
D. four
E. all of the above

_______ 34. Nonferrous metals include _______.
A. aluminum
B. copper
C. brass
D. bronze
E. all of the above

_______ 35. When welding _______, a respirator is required to prevent injury to the welder from toxic fumes.
A. aluminum
B. steel
C. brass
D. all of the above
E. none of the above

SECTION EXAM 4

Name ____________________

Class ____________ Date ____________

Score ____________

Chapters 24-26 Gas Shielded-Arc Welding

True-False

T F **1.** The two general types of gas shielded-arc welding are gas metal-arc welding (GMAW) and gas tungsten-arc welding (GTAW).

T F **2.** Gas tungsten-arc welding uses a consumable steel electrode.

T F **3.** Gas tungsten-arc welding is sometimes called TIG welding.

T F **4.** During the welding operation, a shield of inert gas expels air from the weld area, which contaminates the weld.

T F **5.** Flux is not required when using the gas metal-arc or the gas tungsten-arc welding process.

T F **6.** DCSP is commonly used for welding mild steel.

T F **7.** The direct current reverse polarity side of the ACHF cycle provides deep penetration.

T F **8.** GTAW torches used for welding over 200 amps should be water cooled.

T F **9.** A spherical shaped electrode is used when welding with AC in the GTAW process.

T F **10.** Tungsten electrodes are alloyed with thorium to increase the current capacity.

T F **11.** Argon is commonly used as a shielding gas for gas tungsten-arc welding.

T F **12.** The gas cup size used is determined by the diameter of the tungsten electrode required.

T F **13.** When AC welding, the electrode should touch the base metal when striking an arc with the GTAW torch.

T F **14.** In the GTAW process, filler rod is added to the puddle with an in-and-out motion.

T F **15.** The torch is held at a 45° angle when welding a T-joint in horizontal position.

T F **16.** A gas lens is used to control the arc at the torch.

T F **17.** Three main types of metal transfer can be used in the GMAW process.

T F **18.** DCRP is commonly used in the GMAW process.

T F **19.** Helium with oxygen is used as a shielding gas for welding mild steel in the GTAW process.

T F **20.** Amperage can be adjusted on the power supply in the GMAW process.

T F **21.** The globular type transfer is used for welding thin gauge metal.

T F **22.** A push angle or drag angle can be used in the GMAW process.

T F **23.** The electrode extension used depends on the type of joint being welded.

T F **24.** Gas drift can cause weld defects by allowing nitrogen and oxygen to come in contact with the weld area.

T F **25.** The spray arc type of transfer is used for metals less than $^1/_{16}$″ thick.

T F **26.** Carbon dioxide is used as a shielding gas in the FCAW-S process.

T F **27.** The submerged-arc welding process uses granular flux to protect the weld area from atmospheric contamination.

T F **28.** Flux cored-arc welding most commonly uses DC reverse polarity.

T F **29.** In the gas metal-arc welding pulsed arc process, two current levels are used to obtain maximum penetration without heat buildup.

T F **30.** The submerged-arc welding process is used primarily for welding thin metals.

Multiple Choice

________ 1. In the ________ welding process, the operator applies the weld by hand.
A. manual
B. machine
C. semiautomatic
D. automatic
E. none of the above

________ 2. ________ is the shielding gas most commonly used in the GTAW process.
A. Carbon dioxide
B. Helium
C. Hydrogen
D. Oxygen
E. none of the above

________ 3. The ________ controls the rate of flow of shielding gas to the welding torch in cubic feet per hour.
A. cylinder valve
B. regulator
C. flowmeter
D. check valve
E. none of the above

________ 4. ________ refers to the time in which gas continues to flow after the weld has been completed.
A. Cool time
B. Post purge
C. Preshield
D. Inert flow
E. none of the above

________ 5. A ________ angle is commonly used in the GTAW process.
A. drag
B. pull
C. slip
D. push
E. travel

________ 6. The distance wire projects from the nozzle of a GMAW gun is ________.
A. arc length
B. whiskers
C. cold lap
D. stickout
E. none of the above

________ 7. ________ is recommended when welding aluminum using the GTAW process.
A. DCRP
B. DCSP
C. AC
D. ACHF
E. all of the above

________ 8. ________ added to argon, when used as a shielding gas, increases the penetration achieved in the GTAW process.
A. Oxygen
B. Carbon dioxide
C. Hydrogen
D. Helium
E. none of the above

_______ **9.** The operator can vary the welding amperage during the welding operation with a(n) _______.
A. power supply
B. wire feeder
C. foot control
D. electrode extension
E. none of the above

_______ **10.** _______ is most commonly used when gas tungsten-arc welding mild steel.
A. DCRP
B. DCSP
C. AC
D. ACHF
E. none of the above

_______ **11.** The _______ should be reduced when overhead welding using the GTAW process.
A. shielding gas
B. electrode extension
C. filler rod length
D. amperage
E. all of the above

_______ **12.** _______ is required when welding high-carbon steels using the GTAW process.
A. Preheating
B. Postweld heat treatment
C. A special welding technique
D. all of the above
E. none of the above

_______ **13.** _______ occurs when the welding current is low or below transition current.
A. Spray transfer
B. Globular transfer
C. Short-circuit transfer
D. High-speed transfer
E. none of the above

_______ **14.** _______ is most practical for welding with wire .045″ or less on thinner metals.
A. Spray transfer
B. Globular transfer
C. Short-circuit transfer
D. High-speed transfer
E. none of the above

_______ **15.** _______ requires a high current density and is practical for heavy gauge metal.
A. Spray transfer
B. Globular transfer
C. Short-circuit transfer
D. High-speed transfer
E. none of the above

_______ **16.** _______ is most frequently used in the GMAW process.
A. DCRP
B. DCSP
C. AC
D. ACHF
E. none of the above

_______ **17.** _______ causes the most severe atmospheric contamination problems.
A. Oxygen
B. Carbon dioxide
C. Argon
D. Hydrogen
E. Nitrogen

_______ **18.** _______ occurs when the arc does not melt the base metal completely and weld metal flows outside of the puddle.
A. Porosity
B. Cold lap
C. Burn-through
D. Whiskers
E. Excessive penetration

_______ **19.** The GMAW process _______.
A. does not require electrode change
B. requires a shielding gas
C. does not require slag
D. is faster than the SMAW process
E. all of the above

_______ **20.** A _______ welding machine is used in the GMAW process.
A. constant current
B. constant potential
C. transformer
D. high voltage
E. all of the above

_______ **21.** The _______ is controlled by the wire feeder.
A. amount of amperage
B. voltage
C. type of shielding gas
D. welding gun
E. none of the above

_______ **22.** As the arc length _______, current _______ when using a constant potential welding machine.
A. increases, decreases
B. decreases, increases
C. is lengthened, is adjusted downward
D. all of the above
E. none of the above

_______ **23.** The _______ process uses a tubular wire.
A. GMAW
B. FCAW
C. SMAW
D. GTAW
E. none of the above

_______ **24.** The _______ process does not require a shielding gas.
A. FCAW-S
B. FCAW
C. GMAW
D. GTAW

_______ **25.** The gas metal-arc—pulsed-arc welding process _______.
A. produces a higher ratio of heat input to metal deposition
B. uses different types of metal transfer
C. produces welds with less spatter than welds from GMAW
D. uses pulse peak and background current
E. all of the above

SECTION EXAM 5

Name ______________________

Class ____________ Date ____________

Score ____________

Chapters 27-33 Special Welding Processes

True-False

T F **1.** Brazing can be used to join some dissimilar metals that cannot be joined by welding.

T F **2.** The joints used for brazing are the same joints used for welding.

T F **3.** One factor contributing to the strength of the brazed joint is the adhesive qualities of the filler metal used.

T F **4.** The surfaces to be brazed must be cleaned thoroughly to obtain the necessary flow of filler metal.

T F **5.** The brazed joint is stronger than a welded joint but weaker than a soldered joint.

T F **6.** An oxidizing flame is used for most brazing applications.

T F **7.** The amount of tin and lead in solder determines the melting point.

T F **8.** In braze welding, the filler metal is deposited in a manner similar to that in fusion welding.

T F **9.** In brazing, the base metal must be melted to obtain the necessary penetration.

T F **10.** Flux must be used when brazing and soldering.

T F **11.** Metal is deposited more quickly in hardfacing than in metallizing.

T F **12.** Corrosion is the surface destruction caused by rusting, chemicals, or scaling.

T F **13.** Preheating and postweld heat treatment are necessary when surfacing operations are done on high-carbon steels.

T F **14.** The SMAW process is used extensively for hardfacing because of its high deposition rate.

T F **15.** Metallizing deposits fine semimolten metal particles or metal powder onto the surface of the metal.

T F **16.** The Rockwell hardness tester can be used to determine the bending strength of a particular metal.

T F **17.** Thick wall pipe is $^5/_{16}$″ or greater in wall thickness.

T F **18.** Vertical up welding permits greater penetration and is used on thick wall pipe.

T F **19.** The stringer pass should have a $^5/_{16}$″ crown on the inside of the pipe.

T F **20.** The downhill welding technique is a slower method compared to the uphill technique.

T F **21.** Fast-freeze electrodes are commonly used for welding pipe.

T F **22.** The cap pass reinforces the pipe weld and gives it a neat appearance.

T F **23.** Six tack welds are deposited on the pipe joint to ensure proper root opening and alignment.

T F **24.** Mapp, natural, propane, or acetylene gas with oxygen can be used to provide the necessary flame for flame cutting.

T F **25.** For best results, the preheat and cutting flame must be neutral when flame cutting.

T F **26.** The air carbon-arc cutting torch uses carbon electrodes and compressed air.

T F **27.** The air jet orifices must be positioned above the electrode for maximum metal removal.

T F **28.** Preheat holes are located around the oxygen cutting hole of the flame cutting torch tip.

T F **29.** Inertia welding is a form of resistance welding used in production settings.

T F **30.** Carbon dioxide is used as plasma gas in plasma-arc welding.

T F **31.** Roller type electrodes are used in projection welding.

T F **32.** Gas tungsten-arc spot welding produces a weld with deeper penetration than conventional spot welding.

T F **33.** Ultrasonic welding uses vibratory energy to disperse moisture, oxides, and irregularities between the weld pieces.

T F **34.** The beam-in-air electron beam welding process requires a vacuum chamber to prevent atmospheric contamination in the weld area.

T F **35.** The type of filler rod used for plastic welding is determined by the properties of the base material and the type of joint being welded.

T F **36.** Thermosetting plastics can be welded using the hot-gas method.

T F **37.** A welding gun with a heating element, and compressed gas are used in hot-gas welding to soften the base metal and filler rod.

T F **38.** Two types of robots commonly used are the rectilinear and the articulating.

T F **39.** The robot welder positions the workpiece in a predetermined location.

T F **40.** The robot controller is used to program specific weld functions.

Multiple Choice

________ **1.** The heat used for ________ brazing is generated in the same manner as it is for spot welding.
- A. resistance
- B. dip
- C. manual
- D. furnace
- E. all of the above

________ **2.** The filler metal is drawn into the brazed joint by ________ action.
- A. flow
- B. voltage
- C. capillary
- D. flux
- E. none of the above

________ **3.** ________ temperature is the highest temperature a base metal can reach and still remain in a solid state.
- A. Liquidus
- B. Moltenus
- C. Solidus
- D. Metalus
- E. none of the above

________ **4.** Of the types of solders listed below, ________ has the highest melting point.
- A. 50, 50
- B. 70, 30
- C. 30, 70
- D. 60, 40
- E. none of the above

________ **5.** ________ soldering joins two metal pieces without any solder being visible.
- A. Seam
- B. Torch
- C. Furnace
- D. Sweat
- E. none of the above

______ **6.** In brazing, ______ is used to prevent oxides from forming when heat is applied to the base metal.
- A. filler rod
- B. flux
- C. capillary action
- D. sal ammoniac
- E. none of the above

______ **7.** The metal to be brazed should be preheated until a ______ is reached.
- A. dull red
- B. bright red
- C. melting point
- D. all of the above
- E. none of the above

______ **8.** Fluxes for brazing come in ______ form.
- A. paste
- B. liquid
- C. powder
- D. all of the above
- E. none of the above

______ **9.** Of the welding, soldering, and brazing methods of joining metals, ______ requires the most heat and ______ requires the least heat.
- A. soldering, brazing
- B. brazing, welding
- C. soldering, welding
- D. welding, soldering
- E. welding, brazing

______ **10.** Braze welding should be done in the ______ position.
- A. flat
- B. horizontal
- C. vertical
- D. overhead
- E. none of the above

______ **11.** The ______ method of joining metals requires the least heat.
- A. brazing
- B. welding
- C. soft soldering
- D. hard soldering
- E. braze welding

______ **12.** The ______ type of wear includes grinding, rubbing, or gouging actions.
- A. metallizing
- B. hardfacing
- C. impact
- D. corrosion
- E. abrasion

______ **13.** When hardfacing using the shielded metal-arc process, ______.
- A. use a short arc length
- B. use a high amperage setting
- C. remove rust, scale, or dirt with the heat of the electrode
- D. arrange the work in flat position
- E. all of the above

_______ 14. The _______ process is used when hardfacing parts where heavy deposits are required.
A. oxyacetylene
B. plasma arc
C. metallizing
D. shielded metal-arc
E. submerged-arc

_______ 15. _______ electrodes are often used to deposit base layers for other hardfacing electrodes.
A. Chromium
B. Tungsten
C. Carbide
D. Stainless steel
E. none of the above

_______ 16. The _______ method consists of lining up pipe sections length by length and welding each joint while the pipe remains stationary.
A. roll welding
B. stove pipe
C. backing ring
D. thin wall
E. none of the above

_______ 17. Pipe weld positions are identified as _______ positions.
A. work
B. certifying
C. trade
D. test
E. roll

_______ 18. Consumable insert rings commonly used in pipe welding include _______ rings.
A. Class 1
B. Class 2
C. Class 3
D. Class 4
E. all of the above

_______ 19. Which of the following is true about the vertical up technique used for pipe welding?
A. More penetration is obtained compared to the downhill technique.
B. It is used for thin-wall pipe.
C. The weld is started at the twelve o'clock position.
D. It is faster than the vertical down technique.
E. all of the above

_______ 20. Steel is rapidly oxidized into _______ (iron oxide) in the flame cutting process.
A. flux
B. scale
C. slag
D. rust
E. none of the above

_______ 21. The air carbon-arc cutting process uses a _______ electrode.
A. mild steel
B. carbide
C. tungsten
D. carbon-graphite
E. none of the above

_______ **22.** _______ welding is sometimes called friction welding.
A. Resistance
B. Laser
C. Flash
D. Butt
E. none of the above

_______ **23.** _______ welding uses a man-made ruby to create an intense beam.
A. Electron beam
B. Plasma-arc
C. Laser
D. Projection
E. none of the above

_______ **24.** Resistance welding includes _______ welding.
A. seam
B. butt
C. spot
D. flash
E. all of the above

_______ **25.** The plasma-arc welding process commonly uses _______ as a plasma gas.
A. carbon dioxide
B. helium
C. argon
D. hydrogen
E. none of the above

_______ **26.** _______ welding is used for electrical components, splicing of metallic foils, and hermetic sealing of materials and devices.
A. Electro-gas
B. Plasma-arc
C. Inertia
D. Ultrasonic
E. all of the above

_______ **27.** _______ is a thermoplastic.
A. Acrylic
B. Vinyl
C. Polyvinyl chloride
D. Polystyrene
E. all of the above

_______ **28.** When using the hot gas welding technique on plastics, the _______ must meet manufacturer's specifications.
A. element (watts used)
B. air or gas pressure
C. gas
D. temperature
E. all of the above

_______ **29.** In the _______ method used for welding plastic, current flowing through a metal insert creates heat for fusion.
A. friction
B. hot gas
C. induction
D. heated-tool
E. none of the above

_______ 30. In the _______ method used for welding plastic, the edges to be joined are heated to fusing temperature and brought in contact.
A. friction
B. hot gas
C. induction
D. heated-tool
E. none of the above

_______ 31. V-butt joints used in plastic welding are usually beveled at _______°.
A. 90
B. 75
C. 60
D. 45
E. none of the above

_______ 32. Filler rods used for plastic welding are _______.
A. round
B. flat
C. triangular
D. all of the above
E. none of the above

_______ 33. The _______ of a robot is its maximum range of movement capability.
A. axis
B. rectilinear
C. positioner
D. working volume
E. all of the above

_______ 34. A robot is sometimes called a _______.
A. controller
B. manipulator
C. positioner
D. operator
E. none of the above

_______ 35. The _______ type of robot is particularly suited for intricate maneuvering around the work.
A. rectilinear
B. articulating
C. positioner
D. control
E. none of the above

_______ 36. Robotic welding is a(n) _______ type of welding operation.
A. manual
B. semiautomatic
C. machine
D. automatic
E. all of the above

SECTION EXAM 6

Name ______________________

Class ____________ Date ____________

Score ____________

Chapters 34–36 Supplementary Welding Data

True-False

T F **1.** The dye penetrant method of locating weld defects uses dyes suspended in liquid.

T F **2.** Ammonium persulfate, hydrochloric acid, and nitric acid are used in the etch test.

T F **3.** Destructive tests used include the eddy current, radiographic, and ultrasonic tests.

T F **4.** A tensile test involves bending the metal into a U shape to examine the fusion of the weld.

T F **5.** The percentage of elongation indicates the plasticity of the weld.

T F **6.** Internal defects are located using the destructive and nondestructive methods.

T F **7.** A guided bend test is used to check the quality of fusion and the degree of weld penetration.

T F **8.** The magnetic particle inspection method is used on metals that magnetize and contain iron.

T F **9.** Two specimens prepared for the impact test are the free bend and shear specimens.

T F **10.** The Rockwell hardness tester is used to determine the tensile strength of a weld area.

T F **11.** The five basic weld joints include the butt, T, lap, edge, and fillet.

T F **12.** The two sides of the weld symbol indicate the arrow side and the other side.

T F **13.** The symbol for a fillet weld is a triangle.

T F **14.** Seam welds are dimensioned either by size or strength.

T F **15.** Combined weld symbols are positioned on top of each other on the reference line of the weld symbol.

T F **16.** Plug welds are dimensioned by the size of the leg.

T F **17.** Size is not shown for symmetrical groove welds with complete penetration.

T F **18.** Flange weld size is determined by the bevel angle required.

T F **19.** The weld-all-around symbol is indicated by a small circle located where the arrow line connects the reference line.

T F **20.** The reference tail is omitted unless it includes information about the weld that must be called out.

T F **21.** Test specimens prepared for certification tests must meet the exact specifications required by the certifying agency.

T F **22.** Weld profiles are evaluated on fillet welds and butt welds.

T F **23.** The minimum fillet size for qualification requirements for structural welding is $^{1}/_{8}$″ on base metals with a thickness of less than $^{1}/_{4}$″.

T F **24.** Test specimens used for testing pipe and tubing include the guided bend test and the tensile test.

T F **25.** Specifications of fabrication procedures include location of weld, process, and method of testing.

T F **26.** The American Welding Society is the only agency which has established codes for welding.

T F **27.** A welder must be certified for each welding assignment by the employer or a designated certifying agent.

T F **28.** The 6G test position is for pipe and tubing systems.

T F **29.** AR-1 is the qualification requirement level that applies to systems where a nominal degree of weld quality is required.

T F **30.** The 3G test position for pipe and tubing systems specifies testing in the horizontal position with the axis of the pipe in the vertical position.

Multiple Choice

______ **1.** The ______ test is a nondestructive test.
- A. etch
- B. impact
- C. tensile
- D. eddy current
- E. all of the above

______ **2.** The ______ test is a destructive test.
- A. visual inspection
- B. magnetic particle
- C. shear
- D. radiographic
- E. none of the above

______ **3.** ______ can be checked using the visual inspection method.
- A. Grain growth
- B. Overlaps
- C. Tensile strength
- D. Internal weld defects
- E. all of the above

______ **4.** The ______ test is used to check root bend and face bend test specimens.
- A. impact
- B. tensile
- C. shear
- D. guided bend
- E. all of the above

______ **5.** The ______ test is used to measure the pulling force required to break the weld.
- A. guided bend
- B. free bend
- C. nick-break
- D. tensile
- E. none of the above

______ **6.** The weld specimen must be filed and polished with an abrasive cloth when using the ______ test.
- A. dye penetrant
- B. eddy current
- C. etch
- D. ultrasonic
- E. none of the above

______ **7.** The ______ test cannot be used on nonferrous metals.
- A. magnetic particle
- B. dye penetrant
- C. ultrasonic
- D. radiographic
- E. none of the above

______ **8.** ______ are a set of regulations covering permissible materials, service limitations, fabrication, inspection, testing procedures, and qualifications of welding operators.
A. Standards
B. Specifications
C. Certifications
D. Codes
E. none of the above

______ **9.** ______ are specific regulations which cover the quality of a particular product to be fabricated by welding.
A. Certifications
B. Specifications
C. Requirements
D. Standards
E. none of the above.

______ **10.** The ______ is a test specimen removed from the welded test pipe or tubing.
A. face bend
B. root bend
C. tensile
D. all of the above
E. none of the above

Matching

Identify the weld symbols shown.

______ **1.** Fillet
______ **2.** Plug or slot
______ **3.** Spot or projection
______ **4.** Seam
______ **5.** Back or backing

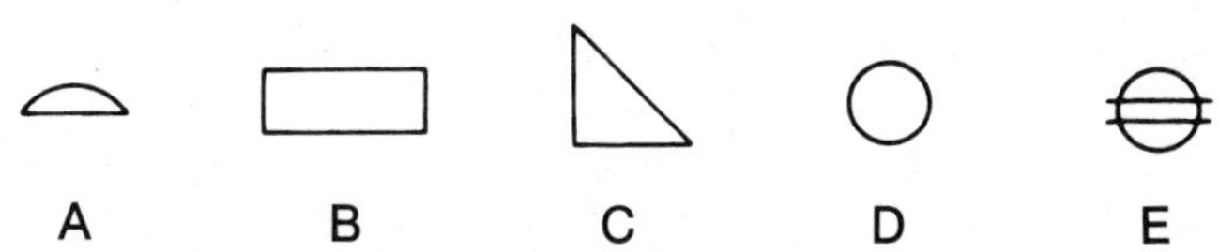

______ **6.** Melt thru
______ **7.** Surfacing
______ **8.** Edge flange
______ **9.** Corner flange
______ **10.** Scarf for brazed joint

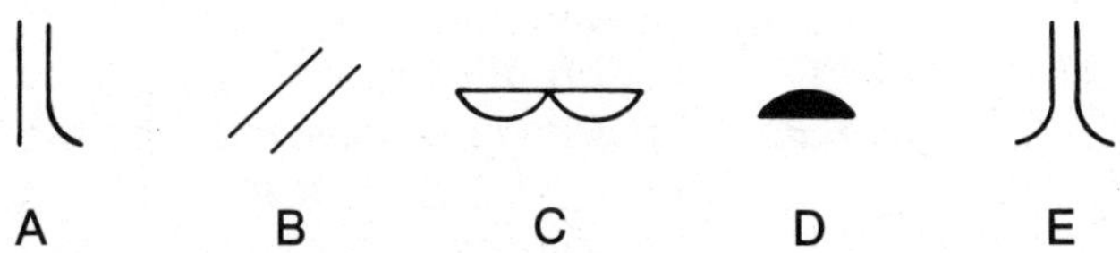

______ **11.** Square-groove
______ **12.** V-groove
______ **13.** Bevel-groove
______ **14.** U-groove
______ **15.** J-groove

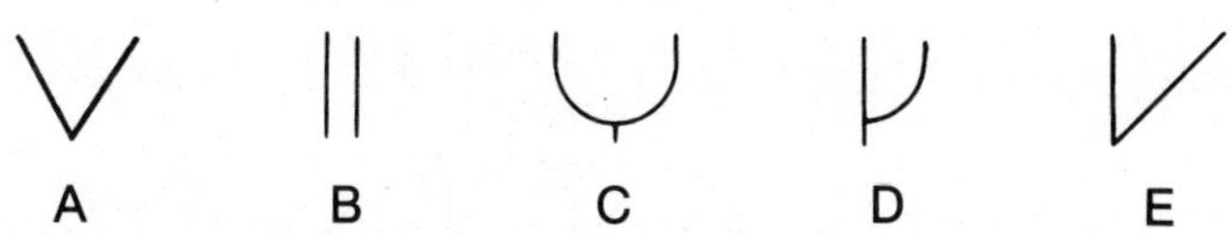

__________ **16.** Flare-V

__________ **17.** Flare-Bevel

__________ **18.** Weld around

__________ **19.** Field weld

__________ **20.** Flush contour

A

B

C

D

E

__________ **21.** Convex contour

__________ **22.** Concave contour

__________ **23.** Reference tail

__________ **24.** Intermittent weld

__________ **25.** Effective throat

A

B

C

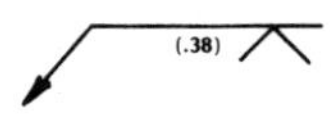

D

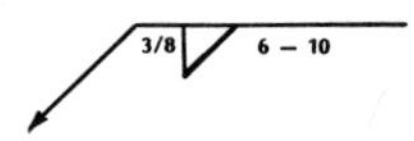

E